Figueroa Florez
Ciro Velasquez

Hydrocolloids in the colloidal stability of fruit drinks

Figueroa Florez
Ciro Velasquez

Hydrocolloids in the colloidal stability of fruit drinks

Incorporation of hydrocolloids in tree tomato and aloe vera beverages

ScienciaScripts

Imprint
Any brand names and product names mentioned in this book are subject to trademark, brand or patent protection and are trademarks or registered trademarks of their respective holders. The use of brand names, product names, common names, trade names, product descriptions etc. even without a particular marking in this work is in no way to be construed to mean that such names may be regarded as unrestricted in respect of trademark and brand protection legislation and could thus be used by anyone.

Cover image: www.ingimage.com

This book is a translation from the original published under ISBN 978-613-9-01035-6.

Publisher:
Sciencia Scripts
is a trademark of
Dodo Books Indian Ocean Ltd. and OmniScriptum S.R.L publishing group

120 High Road, East Finchley, London, N2 9ED, United Kingdom
Str. Armeneasca 28/1, office 1, Chisinau MD-2012, Republic of Moldova, Europe
Printed at: see last page
ISBN: 978-620-7-99278-2

Contents

Summary

The stability of food suspensions is an important factor during development, processing and marketing in order to offer a quality product to the consumer. The aim of this study was to evaluate the effect of hydrocolloids and aloe vera gel on physicochemical, sensory, rheological and stability properties of tree tomato functional beverages. High concentrations of hydrocolloids controlled the phase separation of the beverages during storage. Concentrations of xanthan gum and CMC at 0.05% were found to be adequate treatments in the physical stability of the beverages, expressed in low settling velocities and high Z-potential values (>30mV), without significantly affecting physicochemical properties and colour parameters. The incorporation of aloe vera gel significantly affected pH and titratable acidity ($p>0.05$), but its effect on physical instability was not significant. The rheological characteristics of the beverages were evaluated by rotational and oscillatory tests. The results showed that for the range of concentrations and temperatures studied, the formulated beverages show a pseudoplastic behaviour where the Power Law model showed the best fit to the experimental data. Oscillatory tests reveal the predominance of the elastic modulus ($G'>G''$) over the whole frequency range. The concentrations of xanthan gum ≥ 0.025% and CMC ≥ 0.05% are evaluated as the best treatments, associated to the increase of the consistency coefficient (K) and the trend of the loss tangent (δ) values, which are assumed to be excellent indicators of stability in suspensions.

Keywords: Hydrocolloids, sedimentation, Z-potential, stability, pseudoplastic.

Introduction

Tree tomato (*Cyphomandra betacea)* is a crop of great socioeconomic importance for the Colombian Andean region, as it is a source of livelihood for small and medium-sized producers. It has become a promising exportable fruit, and the orange variety has been the most internationally accepted, due to its colour and nutritional characteristics (Márquez et al., 2007). In Colombia, tree tomato cultivation occupies an area of 8,399 hectares, and the department of Antioquia participates with a production of 2,212 hectares with a yield of 33.6 t/ha (MADR-EVA, 2015). However, despite being a low-cost product, substantial post-harvest losses are reported, affecting the income level of producers. Fresh presentation is the main source of commercialisation, and the lack of processing alternatives hinders the generation of added value and the use of raw materials classified as second or third quality.

The incorporation of tropical fruits in juice production is an alternative to reduce post-harvest losses, reduce production surpluses and add value to the product. A fruit juice is considered a multiphase system determined by a continuous phase consisting of a solution of sugars, salts and other solutes; and a dispersed phase consisting of other solutes such as proteins, fats, vitamins, among others (Duque *et al.,* 2011). When these compounds are reconstituted in water, they become a complex system where dispersions are formed that affect the stability conditions and sensory characteristics of beverages, associated with instability mechanisms such as sedimentation or flocculation, reducing the degree of consumer acceptance of the product (Genovese et al., 1997). Studies highlight the functional and nutraceutical properties of Aloe vera gel (*Aloe barbadensis* Miller), but its inclusion in food suspensions can affect physical stability, given that it is a mucilaginous gel made up of carbohydrates, amino acids, lipids, sterols, minerals and vitamins (Choi and Chung, 2003; Hamman, 2008; Domínguez et al., 2012).

Hydrocolloids are polysaccharides or high molecular weight substances, used in food suspensions as stabilisers, because they increase the viscosity of the continuous phase and lead to ionisation of particles in aqueous solutions (Genovese and Lozano, 2006). Since particles in fruit drinks are negatively charged, the addition of anionic hydrocolloids is expected to increase the electrostatic repulsion forces between particles. Xanthan gum and Carboxymethylcellulose (CMC) are negatively charged hydrophilic polysaccharides, leading to spherical and electrostatic stabilisation of insoluble particles (Genovese and Lozano, 2001; Liang et al., 2006). Although they are used in small concentrations (< 1.0%), they have a high water retention capacity, regulating the rheological and textural characteristics of food systems, creating a gel-like structure and providing physical stability (Sahin and Ozdemir, 2007).

Therefore, the objective of this research was to develop a functional tree tomato beverage by evaluating the effect of the addition of hydrocolloids (xanthan gum and sodium carboxymethylcellulose) and aloe vera gel on physicochemical, sensory, rheological and stability properties. For this purpose, a methodology defined in three phases was established: Phase I corresponding to the formulation and physicochemical and sensory characterisation of the tree tomato drinks; Phase II corresponding to the evaluation of the degree of stability of the formulated drinks; and Phase III consisting of the rheological characterisation of the drinks, based on rotational and oscillatory tests, as a function of temperature. A central rotational design with four central points was established for statistical interpretation and analysis.

The aim of this research is to develop functional beverages as an option for the agro-industrial use of tree tomatoes, to add value to the product and to strengthen the production chain. It is of interest to pursue the implementation of technologies for the elaboration of a functional beverage, which can be scaled up to the public or private sector. In addition, it is intended to offer a good quality product incorporating aloe vera gel as a physiologically active component, enhancing the application of the concepts of healthy and safe food. As a result of this research, it is expected to have a significant effect in the field of Science, Technology and Innovation, through the development of a new alternative for the production of functional tree tomato drinks, a product that does not currently exist and that can be potentialised as an innovator in the agro-industrial market, benefiting both producers and consumers, highlighting that it is a fruit that is highly consumed at a domestic level. It also aims to consolidate the formation of knowledge in the area of rheology and stability of suspensions of a biological nature, an area of research that at headquarters level is opening up space in the agri-food area, significant and necessary information to consolidate technology transfer to agro-industries producing and processing fruit.

In Colombia, not enough research has been carried out on the stability of functional fruit juice drinks using hydrophilic colloids. Therefore, the aim is to lay the foundations to solve a problem of interest in the food sector derived from the colloidal instability that occurs in the development of fruit pulp-based beverages, which reduces the sensory and commercial quality of the product, affecting the marketing process. The generation of new knowledge specifically in the area of suspension stability will contribute to the training of professionals in the area with a view to reducing the existing technological gaps in issues related to the handling, characterisation and development of solutions, suspensions and emulsions of an agri-food nature.

1 Chapter

Theoretical Framework

1.1 Tree tomato: General

The tree tomato *(Cyphomandra betacea)* or tamarillo, is a plant belonging to the Solanaceae family. The centre of origin is in South America, where most of the *Cyphomandra* species are native. It is cultivated in regions with not too cold climates, characterised by temperatures between 14 and 20° C, with best yields between 15 and 18° C. It can be grown at altitudes ranging from 1600 to 2400 m above sea level (Restrepo and Chavarriaga, 2007).

It is an exotic fruit with a delicious flavour and aroma. The fruit is an elongated berry in oval or elliptical shape, pointed at both ends, and crowned by a persistent conical calyx and a long pedicel. The size of the ripe fruit ranges from 4 to 10 cm in length and 3 to 5 cm in diameter (Meza and Méndez, 2009). The skin (exocarp) is smooth, and depending on the type of fruit it can be purple, dark red, orange, yellow or red, sometimes with dark longitudinal veining. The colour of the pulp (mesocarp) also varies according to the type, from reddish or orange to yellow or whitish. It has numerous seeds distributed in two loculi and surrounded by black mucilaginous tissue in the purple or red fruits, and yellow in the yellow or orange ones. The skin has a rough texture and an unpleasant taste, the flesh is succulent and somewhat tasteless, and the tissue surrounding the seeds is juicy and sweet and sour (Lucas *et al.,* 2011; Castro, 2013).

Two varieties of tree tomato are produced in Colombia, the giant orange tomato and the giant purple tomato (García, 2008; Osorio *et al.,* 2012). The giant orange tomato is the best known and most commercialised variety. The rind is red-orange when ripe and has vertical greenish-brown stripes. It has an average size of 5.0 cm wide by 8.0 cm long. The flesh is orange and contains about 240 seeds per fruit. In contrast, the fruits of the purple giant variety are oval-shaped, round, the rind is deep purple with faint vertical green stripes, 5.2 cm in diameter and 6 cm in length. The flesh is orange and contains at least 300 seeds per fruit (García, 2008; Sagñay, 2010; Cáceres, 2012).

Nutritionally, it is an important source of carotene (provitamin A), vitamin C (ascorbic acid) and vitamin E (Table 2-1). It has high levels of protein. It is rich in minerals, especially calcium, magnesium, potassium and phosphorus. It also has a low carbohydrate content and less than 40 calories per 100 g (Márquez *etal.,* 2007; Lagos, 2012).

Table 1-1: Chemical and nutritional characterisation of tree tomato (*Cyphomandra betacea)* (Castro, 2013).

Analysis	Giant orange variety	Giant purple variety
Humidity (%)	87,16 ±1,17	89,2 ± 0,22
Ash (%)	0,81 ± 0,03	0,80±0,01
Protein	2,4±0,04	2,2 ± 0,008
e^H .	3,76 ± 0,04	3,45±0,01
Soluble solids (° Brix)	12,7 ±1,0	10,7 ±1,0
Titratable acidity (% citric acid)	1,87 ± 0,04	1,91 ±0,02
Vitamin C (mg/100g)	0,33±0,19	0,28 ± 0,23
Total sugars (%)	8,6±0,08	4,5 ± 0,06
Total polyphenols (mg/g)	0,84 ± 0,01	0,8 ± 0,01
Total carotenoids (mg/g)	0,23±0,01	0,241 ±0,01
Sugars (%) Glucose	1,38±0,03	1,17±0,03
Fructose	1,64±0,10	1,34±0,02
Sucrose	2,21 ± 0,03	1,86±0,01
Organic acids (mg/g) Citric acid	7,22 ± 0,23	9,19±0,31
Malic acid	1,22 ± 0,04	Not detected

Minerals (µg/g) Calcium	90±1,0	86±1,0
Magnesium	1284	1403
Potassium	3852	3733
Phosphorus	347	281
Sodium	16±1,0	32±1,0
Iron	3 ± 0,04	4 ± 0,04
Zinc	2 ± 0,02	2 ± 0,02

It is low in fat and calories, is a good source of pectins and provides significant amounts of bioactive compounds such as anthocyanins, carotenoids and aavonoids (Castro *et al.,* 2013). These compounds are not only responsible for the colour of the fruit, they also possess biological, therapeutic and preventive properties. The antioxidant activity of different tree tomato extracts has been evaluated, where the results show a high radical scavenging potential by DPPH and ABTS methods (Kou *et al.*, 2009; Ordoñez *et al.,* 2010). This activity is probably associated with the compounds mentioned above. Also, Castro *et al.,* (2013) cite that tree tomato epicarp is a promising and important source of natural antioxidants, which have a potent inhibitory effect on lipid oxidation. Other chemical studies of the fresh fruit indicate that it contains few carbohydrates. The ripe fruit has less than 1% starch and 5% sugars such as sucrose, glucose and fructose. The identification of galacto-arabino-glucuronoxylan, a polysaccharide isolated from the edible pulp of the tree tomato, has also been reported, which has been attributed with analgesic and anti-inflammatory effects (Nascimento *et al.,* 2013). On the other hand, the presence of spheroidal alkaloids such as spirosolanes, solasodine and tomatidenol has been reported, being the ones that have received most attention as alternative sources of spheroids of pharmaceutical interest (Calvo, 2009).

The use of the fruit varies, it is not only consumed fresh (with or without seeds), as a dessert fruit or in salads, but it is also considered a raw material for the preparation of jellies, jams, concentrated or clarified juices, nectars, pulps, preserves in syrup, ice cream, baked goods, dehydrated flakes with fructose, among other uses (León *et al.,* 2004; Castro, 2013; Brito *et al.,* 2008; Jibaja, 2010). However, for consumption, the epicarp is generally removed (due to its bitter, acidic and astringent taste), obtaining a pulp yield of 83-86%. This residue is not really used and in many cases is a source of contamination, despite being considered a promising source of antioxidants that can be used in food preservation (Farfán and Zambrano, 2010; Castro *et al.,* 2013).

In medicine, it is prescribed as an adjuvant in treatment to strengthen the brain, as it helps to cure migraines and severe headaches. Studies to date indicate that it contains substances such as gamma-amino-butyric acid which lowers blood pressure. As a beverage, it benefits the circulatory system by being used in weight reduction programmes due to its satiating effect (Sagñay, 2010). It is also used to combat inflammation of tonsils or angina topically, liver problems, flu and cholesterol control (León *etal.,* 2004; Amaya *etal.,* 2006; Lucas *etal.,* 2011).

From a physiological point of view, the tree tomato is considered a biological system that breathes, transpires and releases ethylene. During its development, it shows a simple sigmoid growth curve and during its maturation it behaves as a non-climacteric fruit. Once harvested, it shows a series of physicochemical, sensorial and biochemical changes, among others; for this reason, the right moment of maturity is one of the aspects that directly affect the post-harvest life and its commercialisation. Between 120 and 150 days of fruit development, the purple colour gradually replaces the green one. Inside, the flesh changes to orange and the stalk loses flexibility. The major changes in acidity, astringency and sugars occur between 150-180 days. The fruit can be harvested at 120 days of development, however, the maximum accumulation of dry matter is reached at 140 days (García and

García, 2001; Márquez *etal.,* 2007; García, 2008).

NTC 4105/97 dictates some commercial quality criteria based on physicochemical indices (determination of diameter, colour, pulp content, consistency, total soluble solids, among others), where it is recommended to harvest the fruit between stages 3 and 4, and keep the stalk to provide protection against fungal and bacterial attack (García, 2008; Castro, 2013). In the post-harvest stage, the most favourable conditions for storage are at temperatures between 3.0 and 4.5° C and relative humidity between 90-95%, temperatures below 3° C cause chilling damage and above this range fungal damage increases. In refrigerated storage at 7° C, using controlled atmospheres with oxygen (O2) and carbon dioxide (CO2) concentrations of 3-5%, allow for extended shelf life. Depending on the variety and without refrigeration, the fruit has a shelf life of 14 to 18 days. Under refrigerated conditions the shelf life is considerably extended to 88 days (Black and Ortega, 2005; Márquez *etal.,* 2007; Lucas *etal.,* 2011).

1.2 Tree tomato production in Colombia

In 2013, according to MADR-EVA statistics (2015), Colombia had 8,399 hectares planted with tree tomatoes, for a total production of 163,748 tonnes. During this period 29.17% of the area planted was in Antioquia, 26.17% in Cundinamarca, 8.47% in Tolima and 7.67% in Huila, among others (Table 2-2).

Table 1-2: National tree tomato production for 2013 (MADR-EVA, 2015).

Department	**Production (tonnes, t)**	**Area Planted (Hectares, ha)**	**Yield (t/ha)**
Antioquia	82.391	2.450	33,6
Boyaca	6.543	523	12,5
Caldas	1.807	137	13,2
Cauca	959	92	10,5
Cesar	624	102	6,1
Cundinamarca	42.120	2.198	19,2
Chocó	8	3	2,5
Huila	4.307	644	6,7
La Guajira	534	81	6,6
Magdalena	2.675	279	9,6
Target	784	28	28
Nariño	2.586	381	6,8
Norte de Santander	2.272	204	11,2
Quindio	258	37	7,0
Risaralda	1.621	141	11,5
Santander	992	110	9,0
Tolima	10.905	711	15,3
Cauca Valley	1.575	186	8,5
Putumayo	790	93	8,5
Total	163.748	**8.399**	**19,5**

The cultivation of this fruit is concentrated in Antioquia and the Altiplano Cundiboyacense, regions where its consumption and commercialisation is quite wide. The cultivation of tree tomatoes represents 3.30% of the harvested fruit area at national level. As for its participation in the food basket, the variation of the Consumer Price Index (CPI) for 2013 in the group of fresh fruit, it was noted that its average (13.06%) is above such important fruits as bananas, mangoes, apples and blackberries (MADR-SEA, 2013).

For the same year, the yields presented by the department of Antioquia exceeded the national average yield, reaching an average of 33.6 t/ha compared to the 19.5 t/ha average registered by the national production. In 2013, the share of national tree tomato production in Colombia was concentrated in Antioquia (50.32% of total production), Cundinamarca

(25.72%) and Tolima (6.66%). According to data from MADR- SEA (2013), in 2013 tree tomatoes accounted for 4.46% of fresh fruit production. It ranked fourth after citrus fruits, pineapple and mango, with a production of 163,748 tonnes.

1.3 Aloe vera in the formulation of functional foods

A functional food is a product that, in addition to satisfying basic nutritional needs, can provide health benefits and reduce the risk of contra-diseases (Juárez, 2010). Other authors stress that it can be a natural food, a food to which a component has been added, removed or modified by biotechnological means, a food in which the bioavailability of one or more of its components has been modified, or a combination of these possibilities (Ferrer and Dalmau *et al.,* 2001). If the biologically active components are incorporated in pharmaceutical forms in high doses, they are considered nutritional supplements or nutraceuticals, but not functional foods (Juarez, 2010; Lutz, 2012). The combination of bioactive components with each other or with other substances is what promotes absorption, transport to tissues, metabolism and protective function against disease. There is a set of functional components such as: carotenoids, dietary fibre, fatty acids, aavonoids, phenolic acids, sterols, polyols, prebiotics, probiotics, minerals and vitamins, which are found in foods, and to which a protective or beneficial effect is attributed (Drago *et al.,* 2006; Juárez, 2010).

Aloe vera gel is nowadays one of the products of vegetable origin that has acquired great commercial importance, due to the potential health benefits of its natural components, and its suitability for use in different industries (cosmetics, pharmaceuticals and food). Aloe vera *(Aloe barbadensis* Miller) is a perennial succulent plant belonging to the *Liliaceae* family. It is similar to a cactus, has turgid, lanceolate leaves, with spines on their serrated edges and bright green, arranged in a rosette. Most used are the leaves, each consisting of three layers: la epidermis or inner layer which is a transparent gel containing 99% water and the rest is made up of glucomannans, amino acids, lipids, sterols and vitamins; la intermediate layer or latex with mucilaginous characteristics which is la bitter yellow sap contains anthraquinones (aloin, aloemodin and phenols) and glycosides; and finally, la thick outer layer called cortex, which has la function of protection and synthesis of carbohydrates and proteins (Domínguez *etal.,* 2012; Khoshgozaran *etal.,* 2012).

Researchers have reported that aloe gel possesses more than 200 different constituents, of which 75 have biological activity (Table 2-3). Studies conclude that the functional properties of aloe are due to the synergy of all the active components and cannot be attributed to the beneficial effect of just one. The broad biological activity and the different uses of aloe have led to the development of several studies to establish relationships between the components and biological effects. Thompson *et al.* (1991) argue that in wound healing, angiogenesis is an essential process. (1998) suggested increased stimulation of pulmonary artery endothelial cells as a result of the angiogenic activity of aqueous aloe vera extracts. Some polysaccharides extracted from the gel (named GAPS-1 and SAPS-1) composed of mannose:glucose:galactose have strong superoxide scavenging activity and moderate hydroxyl radical scavenging and lipid peroxidation inhibition activities (Chun-Hui *et al.,* 2007). Other studies reveal a high antioxidant power of phenolic compounds, Havonoids and ethanolic extracts isolated from aloe vera gel and exocarp (Zheng and Wang, 2001; Hu *et al.,* 2005).

Table 1-3: Components and properties of aloe vera *(Aloe barbadensis* Miller) (Adapted from Choi and Chung, 2003; Hamman, 2008; Domínguez *etal.,* 2012)

Constituents	Active ingredient	Ownership and activity
Sugars	Glucose, fructose, mannose, sucrose, galactose.	AntHnflammatories, immuno-modulators, antiviral action,
Carbohydrates	mannan-oligosaccharides, glucomannan, galactomannan, galactan,	wound healing.

	arabinogalactan, pectic substances, xylan, cellulose.	
Lipids and compounds organic	Steroids (campesterol, cholesterol, β-sitosterol), lupeol, salicylic acid, potassium sorbate, triglycerides, lignin, uric acid, lectin, saponins, gibberellin, triterpenes.	Anti-inflammatory, stimulates la angiogenesis, immuno-modulators.
Amino acids	Alanine, arginine, cysteine, glutamic and aspartic acid, glycine, histidine, eucine, lysine, methionine, phenylalanine, proline, serine, tyrosine and valine.	It provides the building blocks of proteins in muscle tissue production, wound healing, anti-allergic.
Anthraquinones	Aloe emodin, aloectic acid, aloin, anthracin, anthranol, barbaloin, chrysopanic acid, emodin, ethereal oil, cinnamic ester, isobarbaloin acid, resistanol.	Analgesic, antibacterial, antifungal and antiviral activity. They are laxatives.
Enzymes	Alkinase, amylase, catalase, lipase, oxidase-phosphatase-alkaline , Carboxypeptidase, cellulase, peroxidase, cyclooxygenase.	Breaks down food, sugars and fats, aiding digestion and absorption of nutrients. Gastro-protective activity.
Hormones	Auxins, gibberellins.	Wound healing , anti-inflammatory.
Vitamins	A, C, E, B1 (thiamine), B2 (riboflavin), B3 (niacin), B6 (pyridoxine), folic acid, vitamin B12.	Antioxidants (A, C and E), immune modulators, mineral absorption.
Minerals	Calcium, sodium, potassium, magnesium, chromium, copper, manganese, zinc, bolus, nitrogen and selenium.	Essential in the proper functioning of various enzymatic and metabolic systems.

In view of the acemannan and lectin content present in the gel, as activators of macrophages to generate nitric oxide and secrete cytokines, it has been attributed immunostimulatory and immunomodulatory properties (Pugh *et al.,* 2001; Strickland *et al.,* 2001; Im *et al.,* 2005; Broudreau and Beland, 2006). In relation to gastro-protective activity, beneficial effects have been suggested in irritable bowel syndrome, ulcerative colitis and anti-inflammatory activity of intestinal disease in human patients. Inhibition of gastric acid production, stimulation of pepsin and mucosal secretions in rats are other beneficial activities associated with it (Suvitayavat *et al.,* 2004; Langmead *etal.,* 2004; Yusuf *etal.,* 2004; Davis *etal.,* 2006; Pogribna *etal.,* 2008).

The large amount of polysaccharides representing about 20% of the total solids of the mucilaginous parenchyma of aloe leaves are associated with 20 glycoproteins, contributing to the pharmacological activity of the gel as hypoglycemic, hypolipidemic, hepatoprotective, anti-inflammatory, anticancer and skin moisturizing effect (Rajasekaran *et al.,* 2006; Prabjone *et al.,* 2006; Broudreau and Beland, 2006; Steenkamp and Stewart, 2007; Chandan *et al.,* 2007; Hamman, 2008; Kim *et al.,* 2009). Other studies report antimicrobial and antiviral activities of aloe gel associated with anthraquinone content (Rivero *et al.,* 2002; Alves *et al.,* 2004; Habeeb *et al.,* 2007). The potential health benefits found in the natural components of aloe vera have led to its use as a resource in functional foods. Furthermore, highlighting that the inclusion of such components strengthens the nutritional value of foods (Boghani *et al.,* 2012).

Aloe gel finds wide application in food products such as nectars, juice concentrates, health drinks, laxative drinks, yoghurts, ice cream, among others (Eshun and He 2004; Hamman 2008; Singh and Singh, 2009; Ahlawat and Khatkar, 2011). Sierra (2002) evaluated the effect of UV-C light on the inhibition of mesophiles, aerobes, moulds and yeasts in a prototype drink of aloe and orange in equal proportions. Elbandy *et al.* (2014) developed a mango nectar with high therapeutic and nutritional value by supplementing the pulp with aloe gel. The nectar showed excellent quality attributes and good storage stability, showing variations in vitamin C content and physicochemical characteristics such as acidity and pH. Boghani *et al.* (2012) developed a beverage by blending papaya and aloe juice, highlighting differences in physicochemical and sensory quality. They also report good stability in the refrigerated storage period.

Wei *et al. (2004)* prepared a functional beverage from fresh aloe vera leaves with Chinese herbs and roots. They studied the effect of processing conditions such as temperature, pH, sucrose and citric acid concentration, concluding that the physical stability of the beverages is negatively affected as sucrose and citric acid concentration increases. Do-Sang *et al,* (1999) prepared vinegar from aloe vera juice using *Acetobactor sp.* Lee and Mano-Yoon (1997) formulated an aloe vera yoghurt with lactic acid bacteria (*Lactobacillus bulgaricus and Streptococcus thermophilus* strains) and compared it with yoghurt prepared from skimmed milk powder, finding retention in quality characteristics and good physical stability of aloe vera yoghurt stored at 5° C for 15 days, with respect to milk yoghurt.

Other applications of aloe vera gel focus on the postharvest area of fruits and vegetables, specifically in the formulation of edible coatings (Restrepo, 2009; Khoshgozaran *et al.*, 2012; Benítez *et al.,* 2013; Ramírez *et al.,* 2013; Yulianingsih *et al.*, 2013). The authors agree that the coated fruits showed a decrease in respiration rate and control of physicochemical, microbiological and sensory parameters compared to the control fruits. In addition, the shelf life of the product was increased. Sigrid and Vidal (2011) studied the feasibility of developing fresh functional foods by incorporating aloe vera gel into the structural matrix of some vegetables through vacuum impregnation. The product obtained retained the characteristics of fresh food (high water activity), while incorporating beneficial properties of aloe, enhancing the intrinsic functionality of the vegetables under study.

1.4 Hydrocolloids

Hydrocolloids or food gums are water-soluble polysaccharides or proteins of high molecular weight, used in a variety of functions in food systems to increase viscosity, form gel structures, form films, control crystallisation, inhibit syneresis, improve texture, encapsulate flavours and extend physical stability, among others (Williams and Phillips, 2009). Due to their functional properties such as water retention capacity, balance of rheological properties and ionisation of aqueous solutions, they allow controlling the degree of instability of insoluble particles in food suspensions (Genovese and Lozano, 2001).

To ensure stability of food suspensions, such as fruit juices or nectars, during prolonged storage, the use of hydrophilic colloids has been promoted (Table 2-4). In terms of ionisation, hydrocolloids may or may not be electrically charged. Since particles in suspension are negatively charged, the addition of anionic hydrocolloids is expected to increase the electrostatic repulsion forces between particles. Also, the adsorption of rubber macromolecules on the particles can lead to spherical repulsion (Genovese and Lozano, 2001). Sahin and Ozdemir, (2007) argue that food gums improve the texture and theological properties of suspensions through an increase of viscosity in the continuous phase by decreasing the phase separation process.

Table 1-4: Hydrocolloids as stabilisers in food dispersions

Hydrocolloids (Matrix)	Effects	Reference
Chitosan (Orange juice)	Increase in turbidity and control of phase separation. Change in physicochemical properties (pH, TSS). Significant increase in colour parameters (L*, a*, b*). Pseudoplastic behaviour.	Martín *et al.*, (2009)
Goma persico and	Pseudoplastic fluid, with power law model fit.	Abbasi *etal,* (2013).

tragacanth (mixed orange-milk juice)	Increased differential zeta potential. Spherical and electrostatic stability. A 0.37& rubber mixture was assessed as the best treatment.	
Xanthan gum and CMC. (Apple juice)	Increase in viscosity with the addition of rubber. Pseudoplastic behaviour. Increase of electrostatic repulsion. Increased turbidity of the clarified phase. Improved juice stability with CMC.	Genovese and Lozano, (2001)
Xanthan gum, CMC, pectin (Apple juice)	Good stability, increased viscosity and cloudiness. Decreased sensory attributes, particularly taste.	Ibrahim *etal,* (2011).
Cross-linked and pre-gelatinised starches. (Apple juice)	Unimodal particle size distribution. Thixotropic fluids, increase of theological parameters with the addition of starches. Better stability associated with pre-gelatinised starches.	Meng and Rao (2005)
Xanthan gum, guar gum (apple juice)	Good stability, increased viscosity of the continuous phase. Better stability with xanthan gum. Pseudoplastic behaviour.	Paquet *et al.,* (2014)

Table 2-4: (continued)

Hydrocolloids (Matrix)	Effects	Reference
Xanthan gum, CMC, gelan gum, guar gum. (Carrot juice)	Increase of viscosity and turbidity, with the addition of gums. Control of phase separation. Improved juice stability with gum gelan.	Liang *et al.,* (2006)
Soy protein (Soursop juice)	Increased viscosity, pseudoplastic behaviour, better fit to Cross model. Reduction of sedimentation rate and sedimentation velocity. Good fit to first order kinetic equation.	Fasolin and Cunha (2012).
xanthan gum, CMC and guar (mixed juice, carrot-orange)	Sedimentation control. Increase in viscosity. Change in physico-chemical properties (pH and acidity).	Nwaokoro and Akanbi (2015).
Goma tragacanth (milk drink)	Increase in apparent viscosity. Unimodal size distribution. No effect on mean particle diameters. Change in colour parameters (L*, a*, b*). Improved sensory properties.	Keshtkaran et al., (2013).
Pectin, CMC (Orange drink)	Increase in viscosity and Z potential differential. Decrease in physicochemical properties (pH). Increase in turbidity.	Mirhosseini and Tan, (2010).

Xanthan gum and Carboxymethylcellulose sodium (CMC) are commonly used as thickening agents in the food industry. Xanthan gum is an extracellular polysaccharide secreted by the microorganism *Xanthomonas campestris,* used for its high thickening capacity, excellent stability over a wide pH range, and resistance to enzymatic degradation (Liang *et al.,* 2006). At low temperatures, xanthan gum molecules exist as a double helix, but become a disordered spiral at higher temperatures. The carboxyl groups on the side chains are responsible for the anionic characteristic of this polysaccharide. The pyruvic acid content can vary substantially depending on the *X. campestris* strain, resulting in different viscosities of xanthan solutions (Milani and Maleki, 2012). It exhibits weak gel-like rheological properties, pseudoplastic behaviour, very viscous at low shear rates, but very low viscosity at high shear, and quickly regains its viscosity when mechanical stress is removed. Xanthan gum confers excellent stability of suspended particles during storage and transport, and acquires good flowability during agitation and pumping processes. All these factors make xanthan gum an ideal stabiliser for suspensions, sauces and dressings. Another important use is during baking of bakery products, increasing volume, maintaining moisture and particle suspension in dough (Pegg *etal.,* 2012).

Carboxymethylcellulose sodium (CMC) is a water-soluble anionic polymer capable of forming highly viscous solutions, which gives it excellent properties as a stabiliser. CMC is prepared by first treating cellulose with alkali (alkaline cellulose), and then by reaction with monochloroacetic acid (Milani and Maleki, 2012). Concentration, molecular weight and degree of substitution (GS) are important factors for the flow behaviour of CMC in aqueous dispersions. Commercial products generally have GS values from 0.7 to 1.5 (Saha and Bhattacharya, 2010). The polymer chain exhibits a helical conformation in suspension, which significantly influences its rheological behaviour. CMC gum dispersions exhibit pseudoplastic rheological properties, which are significantly affected with increasing temperature. In addition, weak gel-like viscoelastic behaviour has been confirmed at high concentrations (Benchabane and Bekkour$_1$ 2008; Saha and Bhattacharya, 2010). CMC is widely used as a thickener, water binder, encapsulation and film forming adjuvant in the food and pharmaceutical industry to improve consistency and flow properties (Yasar et al., 2007). It is unusual in its ability to interact with proteins such as casein and soy, and protect them from precipitation at their isoelectric point. This reduces its application in dairy products (Pegg *et al.,* 2012).

The viscosity of xanthan gum solutions is obviously higher than that of CMC at low shear rates (<10s^{-1}), which explains the excellent stabilising properties (Liang *et al.,* 2006). The selection of a hydrocolloid depends on its physicochemical characteristics, as well as its price and safety. Not surprisingly, starches are the most frequently used thickeners due to their low cost resulting from their high annual production rate. However, a more expensive thickener such as xanthan gum may still be the first choice due to its unparalleled rheological properties (Li et al., 2015). Lazaridou et al., (2007) noted that compared to CMC and pectin, xanthan gum possessed lower creep deformation, higher viscosity at low shear rates, increased elastic properties and predominance of elastic modulus giving it the weak gel-like characteristic.

1.5 Suspension stability

A dispersion is a polyphase system in which one phase is fragmented within another. There are three types of dispersed systems called foam, emulsion or suspension. The fragmented phase is a certain amount of gaseous, liquid or solid matter, which is called a bubble, droplet or particle if they are of macroscopic size. A suspension is a colloidal dispersion in which a *solid* is dispersed (internal phase) in a continuous *liquid* phase (external phase). Suspensions can be aqueous or non-aqueous (Perez, 2009). The classical size range of colloidal dispersions is between 1nm - 1mm, and it is assumed that the dispersed species are spherical in shape. When other shapes are considered, particles with diameters up to 2mm can be described as colloids. In practice, suspensions usually have diameters larger than 0.2mm, and sometimes contain particles that exceed the limits of the classical range (5010mm diameter). Particle sizes can also be below the standard size limit mentioned above, and are known as nanoparticle suspensions (Schramm, 2005).

In the stability of suspensions, a stable, homogeneous dispersion without particle aggregation is sought. However, there are three physical phenomena that cause instability of a suspension: sedimentation, aggregation and coalescence (Núñez and Scrofani, 2011). Sedimentation is the result of a difference in density between the dispersed phase and the continuous phase, and produces two separate phases that have different concentrations and viscosities (Aranberri *et al.,* 2006). Particles in suspension settle differently, depending on the characteristics of the particles as well as their concentration, and can therefore be referred to as discrete particle sedimentation, Aoculent particle sedimentation and free-fall particle sedimentation. Discrete particles are those that do not change their characteristics (shape, size, density) during the fall. In free-falling sedimentation (i.e. no interference between particles), particles are supported by hydraulic forces and their fall can be described by Stokes' law (Torres *et al.,* 2008). Stokes' law is a classical modelling (Eq. 2.1) that allows determining the sedimentation velocity for a rigid, smooth, spherical, non-colloidal particle sedimenting in a viscous Newtonian fluid, which was proposed by Stokes in 1850, considering that the laminar flow regime of the fluid over the particle is laminar, with a range of values for the Reynolds number of less than 0.25 (Salinas and Fuentes, 2012).

$$v = \frac{(\rho_p - \rho_f) \cdot D^2 \cdot Z \cdot g}{18 \cdot \mu} \qquad \textbf{(Ec. 2.1)}$$

Where Z is the acceleration factor, ρ_p is the average particle density (kg/m^3), ρ_f is the average fluid density (kg/m^3), D is the particle diameter (m), g is the acceleration of gravity (m/s^2), μ is the dynamic viscosity of the fluid (Pa.s).

Aggregation, on the other hand, occurs when Brownian motion, sedimentation or an agitation system causes two or more dispersed species to tend to agglomerate, possibly joining at some points without any change in the total surface area. In aggregation, the particles retain their identity, but lose their kinetic independence as the aggregates move as a single unit. Aggregation is sometimes referred to as flocculation or coagulation, but, for suspensions, coagulation and flocculation are often taken to represent two different types of aggregation. In this case, coagulation refers to the formation of compact aggregates, while flocculation refers to the formation of a loose network of particles, linked through edge-to-edge and edge-to-face particle interactions (Schramm, 2005). Coalescence is the process when two or more particles coalesce to form a single larger unit, with the elimination of part of the liquid-liquid-liquid interface. This irreversible change would require extra energy input to restore the original particle size distribution (Núñez and Scrofani, 2011).

Phase instability or phase separation is an unfavourable parameter in the commercial quality of suspensions during storage (Sherafati *et al.,* 2013). In fruit juices, particles with positively charged cores (carbohydrates, proteins) coexist surrounded by negatively charged cellulosic material (native pectin). Degradation of pectins would expose the positive nuclei, may lead to aggregation of polyanions and polycations, and eventually to flocculation of particles (Genovese and Lozano, 2001). Particle size distribution is another important factor in the stability of dispersions, as larger sizes can cause precipitation of insoluble material leading to particle sedimentation (Schramm, 2005). In turn (Genovese et al., 1997) argue that fruit pulps possess a large amount of insoluble polymeric materials, which can increase physical instability leading to food suspensions. Several studies have tried to predict phase separation kinetics through quantitative sedimentation tests. The influence of components on the phase separation process in soursop juices was analysed using a first-order kinetic equation, finding a decrease in settling velocity with increasing soluble soy polysaccharides used as stabilisers (Fasolin and Cunha, 2012). Likewise, Kaneiwa *et al.,* (2013)
implemented a first-order kinetic model to study phase separation in tomato juices stabilised at high homogenisation pressures (HPH). They report a reduction in the sedimentation rate, associated with a reduction in suspended particle size.

The stability of a colloidal system is determined by the sum of the electric repulsive double layer forces, and the van der Waals forces of attraction that the particles experience when they approach each other. The repulsive forces must be dominant. There are two general ways in which colloidal stability can be provided: electrostatic and spherical (Genovese *et al.,* 2007). Electrostatic stabilisation results from the presence of a layer that forms or grafts onto the particle surface. Adsorbed surfactants, nanoparticles, or macro-ions provide a similar effect. Repulsion occurs when the adsorbed or grafted layers on two particles begin to overlap and usually increases rapidly as the layers are compressed. The surface of the colloidal particles can acquire a surface charge which in some cases is related to the physicochemical characteristics of the medium, and can be counteracted, with dissolved ions present in the dispersion medium forming an electrical double layer which in turn can interact with the double layers of other particles in suspension, creating a long-range repulsion due to Coulombic-type interactions which can be greater in magnitude than the attraction by dispersion forces. This repulsive interaction leads to stability of the colloidal dispersion and is called electrostatic stability. A greater thickness in the double layer allows for better suspension stability (Urquijo, 2007). Unlike the electrostatic phenomenon, the spherical process can impart true thermodynamic stability, so such dispersions can often be highly concentrated and still remain stable, leading to highly viscoelastic behaviour (Mewis and Wagner, 2012). The use of natural and synthetic polymers to stabilise aqueous colloidal dispersions is technologically important, and extensive research in this area has focused on adsorption and spherical stabilisation. Polymers can cause repulsion between particles, and colloidal spherical stability is achieved when they are embedded in or adsorbed by particles (Schramm, 2005).

The measurement of the potential differential (Z) is used to evaluate the stability of colloidal systems, and is determined by the electrophoresis technique. The square of the potential Z is proportional to the electrostatic repulsive force between charged particles. Each colloid contains an electrical charge, usually of a positive or negative nature. These charges produce electrostatic repulsive forces between adjacent particles. If the charge is sufficiently high, the particles remain discrete, dispersed and in suspension. Removing these charges produces the opposite effect and the colloids agglomerate and settle out of suspension (Pérez, 2009). In the literature, there is a large body of research focused on studying the stability of dispersions through the determination of the Z-potential (Croak and Corredig, 2006; Ly *et al.,* 2008; Benítez et al., 2009; Mirhosseini and Ping, 2010; Chivero *et al.,* 2015). Electrokinetic measurements in oil-water emulsions show that gum arabic has a significant capacity to stabilise the colloidal dispersion. It was also found that gum imparts stability to the emulsion through an electro-spherical mechanism, but the spherical contribution is dominant (Jayme *et al.,* 1999).

Giupponi and Pagonabarraga (2011) performed the determination of the zeta potential for highly charged colloidal suspensions, and showed that the concentration of oppositely charged ions in the suspension has a significant effect on the Z potential. Genovese and Lozano (2001), found that the best stability of apple juice is achieved when using 0.4% w/w xanthan gum (Z-potential > -35mV). Abbasi and Mohammadi (2013) deduced through rheological studies and Z-potential determination that the implementation of persian gum leads to stabilisation of electrostatic and spherical repulsions in orange and milk juices (Z-potential > -25 mV). An increase in zeta potential values when guar gum-xanthan gum combinations were implemented in a typical Iranian beverage (Dooh) led to excellent suspension stability (Fayyaz *et al.,* 2014). A broader view of these stability criteria developed for particulate suspensions is detailed in Table 2-5.

Table 1-5: Stability criteria based on zeta potential (Schramm, 2005).

Stability characteristics	Potential Z (mV)
Maximum agglomeration and precipitation	+3 to zero
Excellent agglomeration and precipitation	-1 a -4
Favourable agglomeration and precipitation	-5a-10
Agglomeration threshold (agglomeration of 2-10 particles)	-11 a-20
Light stability plateau (little agglomeration)	-21 a -30
Moderate stability (No agglomeration)	-31 a -40
Good stability	-41 a -50
Very good stability	-51 a -60
Excellent stability	-61 a -80
Maximum stabilityFor solids	-81 a-100
For emulsions	-81 a-125

1.6 Suspension rheology

1.6.1 Importance of rheology in food suspensions

Rheology is the study of the deformation and flow of matter, and is an instrumental technique widely used in the characterisation of raw materials, and in processing and preservation processes (Stading, 2011). Rheological properties of food fluids are useful during food processing and handling, which involves fluid flow in operations such as pasteurisation, evaporation and dehydration. Also, in quality assessment, structure analysis, equipment design, system and transport requirements (Quek *et al.,* 2013).

The theological study of a particle suspension is a complex function of its physical properties and the processes occurring at the particle scale. The most important factors are concentration, size, shape, interactions and deformability (Mueller *et al.,* 2010). Most studies and theoretical aspects are linked to the rheology of non-food dispersions, systems that feature spherical and rigid particles. However, due to the irregular and deformable character of plant material particles, the rheology of food suspensions is much more complex than that of monodisperse suspensions and solid spheres. In addition, interactions can occur between the particles in suspension and the continuous phase, which further complicates the study of rheological parameters (Mewis and Wagner, 2012; Moelants *et al.,* 2013).

Another interesting aspect of the properties of fluid systems focuses on the study of the stability behaviour of suspensions. Structural instability due to sedimentation has been observed in systems involving solid particles and non-Newtonian fluids. When solid particle suspensions of polymers settle under progressive flow conditions, it has been observed that the suspension structure becomes unstable under certain conditions (Phillips, 2010). When considering a beverage as a liquid fluid with characteristics of a colloidal dispersion, the flow properties are crucial in the study of its behaviour. High viscosity can be a problem that can be treated as a resistance to flow, or it can be a desirable property for the formulation of a dispersion. Stokes' law considers the terminal settling velocity to be inversely proportional to the viscosity of a colloidal dispersion, which has a direct impact on particle settling that is indicative of physical instability (Schramm, 2005). Rheological changes in the continuous phase of a dispersion can lead to increased stability of the suspension for longer periods of time due to retention of the suspended particles. The addition of hydrophilic colloids increases the viscosity of a continuous medium, and helps to keep particles in suspension and avoid precipitation (Genovese and Lozano, 2001).

In food products, rheology provides guidelines for the definition of a set of parameters, which can be correlated with other quality attributes (Garriga, 2002). Rheological characteristics can be of great interest to modify the processing or formulation of a final product so that the texture parameters of the food are within the range considered desirable by consumers (Stokes *et al.,* 2013). The influence of the rheology of a particular food on taste perception can have two main origins: First, a physiological effect due to the proximity of taste and olfactory receptors to the kinaesthetic and thermal receptors in the mouth, as an alteration of the physical state of the material can have an influence on its sensory perception, and an effect related to the bulk properties of the material (e.g. texture or viscosity), as the physical properties of the material can affect the speed and degree to which the sensory stimulus reaches the taste receptors. Second, in fluid systems, where hydrocolloid solutions and dispersions are used as model systems, it has been generally shown that an increase in viscosity can reduce the perception of sweetness (Rao, 2006).

Rheological measurements have also been considered as analytical tools that provide fundamental information on the structural organisation of food, whereby the Theological response of a material depends on its molecular interactions. Thus, the Theological parameters measured in the food provide information regarding the mechanical behaviour of the structure. Raeuber and Nikolaus (1980) pointed out that microscopy describes the visible structure, but the mechanical bonding can only be known intrinsically through Theological parameters which are revealed under stress-strain application. Theological properties are sensitive to variations in molecular structure and are useful in the development of structure-function relationships for polysaccharide systems in solution.

1.6.2 Viscosity of a suspension

The forces of attraction that keep molecules at minute distances from each other, inducing sufficient cohesion in fluids, mean that there are forces that oppose the relative movement of adjacent layers of the fluid. This resistance offered by fluids to displacement is known as viscosity. Since there is a great variety of fluids whose behaviour does not conform to that of a Newtonian fluid, at least in a certain range of stresses, the concept of intrinsic viscosity disappears, and then the term apparent viscosity μ appears, which is defined as (Eq. 2.2) (Garriga, 2002).

$$\mu = \frac{\sigma}{\gamma} \qquad \textbf{(Ec. 2.2)}$$

Where, o^is the shear stress [Pa], and yes the strain rate [s].[1]

The viscosity of a suspension is a function of particle shape and concentration. If miscible additives are added to the continuous phase, the viscosity of the suspension will be affected, however, it is difficult to predict how this will affect its behaviour. For dilute suspensions and spherical particles, the variation in viscosity caused by the presence of the dispersed phase in the continuous phase describes Newtonian behaviour, based on the fact that the dispersed phase is very dilute, i.e. the particles are so far apart that they do not interact with each other (Mueller *et al.,* 2010). When working with concentrated suspensions, the flow lines are modified, producing an increase in energy dissipation and thus increasing viscosity, and the fluid exhibits non-Newtonian behaviour. Likewise, when non-spherical particles are present, the suspension manifests a non-Newtonian character; moreover, as these particles are not rigid, they deform with the flow and can exhibit viscoelastic behaviour (Pabst, 2004).

During processing, transport and storage, food suspensions are subjected to different temperature conditions, which make it essential to study the change of rheological properties as a function of this thermodynamic variable. Liquids have invariant viscosity coefficients, which decrease with temperature. In general, the effect of temperature on the rheology of suspensions has been studied by means of the Arrhenius equation (Eq. 2.3).

$$X = X_o\, e^{\left(E_a/RT\right)} \qquad \textbf{(Ec. 2.3)}$$

Where *X* is any Theological property, X_o is a constant, E_a is the activation energy, *Res* is the gas constant and T'is the temperature in absolute scale.

There have been a number of investigations where a significant decrease in viscosity of fruit juices with increasing processing temperature has been reported (Chin *et al.,* 2009; Vandresen *et al.,* 2009; Ibrahim *et al.,* 2011; Quek *et al.,* 2013; Shamsudin *et al.,* 2013). However, no studies are reported on the effect of temperature on the theological behaviour of tree tomato beverages. Some studies have reported that the estimated models adequately represent the behaviour of rheological parameters, such as the consistency coefficient (*K*) and flow behaviour index *(n),* of fruit juices as a function of temperature using the Arrhenius equation. Kaya and Sözer (2005) suggest that this relationship could be successfully used to estimate the temperature dependence on the rheological behaviour of sugar-rich food fluids and low-concentrated fruit juices (Table 2-6). The activation energy *(E_3)* indicates the sensitivity of viscosity to temperature changes. Higher E_3 values mean that the apparent viscosity is relatively more sensitive to temperature.

Table 1-6: Rheological parameters in some food dispersions.

Product	T(°C)	K (Pa.s)^n	n	K_o (Pa.s)^n	Дo (Pa.s)	E_a (kJ/mol)	Bibliography
Peach purée	25-55	0,3 3,4	0,460,78	-	-	24,0 30,0	Warriorand Alzamora, (1998)
Grape juice concentrate	35-65	0,28 2,53	0,720,85	2,4E-03- 3,5E05	-	13,37 28,59	Arslan, (2003)
Blueberry puree	25-60	0,07 7,20	0,640,49	-	-	10,7 21,7	Nindo eta/., (2007)
Grapefruit juice	6-75	0,004 - 3,76	0,501,05	3,0E-04- 9,9E-08	-	22,12 34,02	Chin et al., (2009)
Carrot juice	8-85	-	-	-	2,1E-02- 2,9E-03	12,83 15,30	Vandresen et al., (2009)
Soursop juice	10-50	0.001 - 5,22	0,40 1,03	5,2E-08- 5,8E-02	-	30,48 - 10,23	Queketa/., (2013)
Pineapple juice	5-25	0,001 - 0,012	0,94 0,98	3,0E-04- 5,0E-04	-	6,80 8,50	Shamsudin et *al.*, (2013)
Kiwi juice	25-65	0,15 - 5,56	0,360,41	-	0,3E-04 - 5,3E-04	24,72 - 34,29	Goula y Adamopulos (2011)
Beetroot juice	30-80	0,007 - 0,98	0,620,99	-	1,2E-02 - 2,4E-03	14,80 25,03	Kumar and Kumar (2015)

1.6.3 Determination of apparent viscosity

One of the most suitable geometries for determinations with non-Newtonian liquids is that of concentric cylinders (Fig. 2-1). In these, the strain rate is identical throughout the sample, provided that extreme effects are minimised; furthermore, they allow controlled modification of the tangential strain rate and operating time (Steffe, 1996). The speed adjustment (up or down) and the reading of the corresponding values must be carried out without stopping the rotation. There are coaxial cylinder viscometers in which the fluid is sheared in the space between two coaxial cylinders of different radius, they can operate in such a way that the moving cylinder is either the inner cylinder (Searle type) or the outer cylinder (Coutte type). The moving cylinder can rotate at a constant rate, with the momentum exerted by the fluid being measured, or a constant momentum can be applied and the angular velocity measured (Mewis and Wagner, 2012).

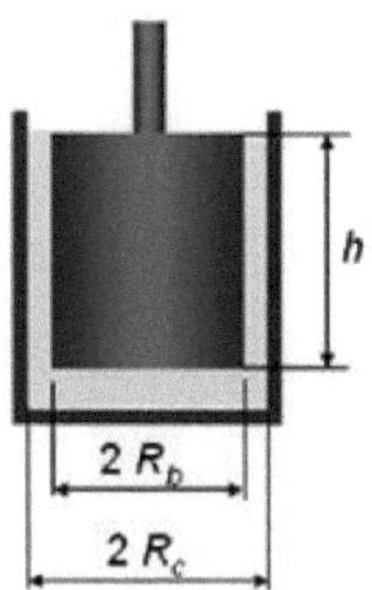

Figure 1-1: Schematic representation of concentric cylinder geometry.
(Zhong and Daubert, (2007).

The main limitations of viscometers with coaxial cylinder geometry are when dispersed systems have phase separations, which cause "slippage" on the cylinder surfaces. In such a case, the equipment registers the viscosity of the continuous phase and not of the homogeneous dispersion. On the other hand, the rotor cylinder must be completely immersed in the fluid so that the forces acting on its ends cancel each other exactly. The closer and more accurate the viscosity determination is, the higher the viscosity of the fluid to be studied (Rao, 2006).

The moving cylinder is rotated at a fixed angular velocity (Ω), and the torque (M) required to keep this shear stress constant is recorded. The angular velocity, expressed in radians/second, is converted to revolutions/minute with the following relationship (Eq. 2.4) (Zhong and Daubert, 2007):

$$1\ \text{rpm} = \left(\frac{1\ \text{revolution}}{1\ \text{min}}\right)\left(\frac{2\pi}{1\ \text{revolution}}\right)\left(\frac{1\text{min}}{60\text{s}}\right) = 0{,}105\ \text{rad/s} \qquad \text{(Ec. 2.4)}$$

For the movable cylinder with radius Rb y a fixed cylinder of radius Re, the shear stress y shear rate, can be estimated as follows (Eq. 2.5, Eq. 2.6).

$$\sigma_b = \frac{M}{2\pi h R_b^2} \qquad \text{(Ec. 2.5)}$$

$$\gamma_b = \frac{\Omega R_b}{R_c - R_b} \qquad \text{(Ec. 2.6)}$$

1.6.4 Rheological models

Numerous rheological models have been used to describe the flow behaviour of foods. These models are used to correlate the behaviour of various fluids over a wide shear range, although sometimes a single model is not sufficient to describe the behaviour of a given fluid. It should be noted that most food fluids do not have Newtonian behaviour (Quek *et al.*, 2013). There are two types of fluids, depending on the application of shear stresses. Time-independent fluids (dilatant, plastic and pseudoplastic), in which the deformation rate is a non-linear mono-valued function of the applied shear stress and have no rheological memory (Garriga, 2002). The most common behaviour in food dispersions is associated with shear-thinning fluids, which are characterised by a decrease in viscosity and shear stress with strain rate (Fig. 2-2). Such fluids exhibit flow behaviour indices of less than unity (^<1.0) (Shamsudin *et al.*, 2013). Table 2-7 lists the most common models used to describe the rheological behaviour of time-independent fluids. The fewer parameters a fluid has, the simpler its rheology and the more applicable it is as a rheometric fitting model (Gratáo *et al*, 2007).

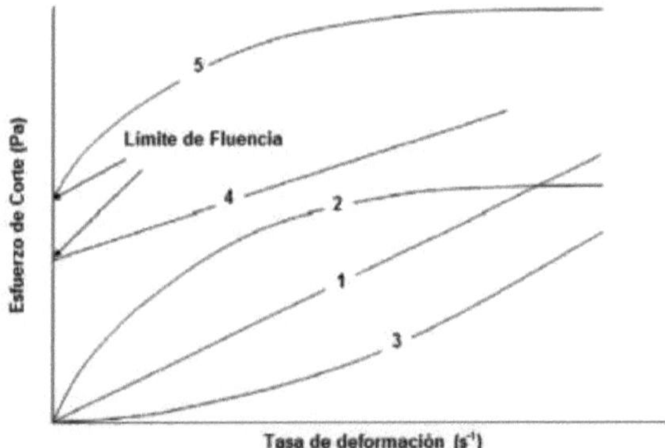

Figure 1-2: Typical behaviour of time-independent fluids: 1) Newtonian, 2) Pseudoplastic, 3) Dilatant, 4) Bingham plastic, 5) Herschel-Bulkley (Zhong and Daubert, 2007).

Table 1-7: Time-independent rheological models of non-Newtonian fluids.

Model	Equation* Equation* Equation* Equation* Equation* Equation*	Parameters

	Equation* Equation* Equation	
Casson	$\sigma^{0.5} = (\sigma_o)^{0.5} + K_1(\gamma^{0.5})$	σ_o, K_1
Mizrahy and Berk		
Ellis	$\sigma^{0.5} = (\sigma_o)^{0.5} + K_1(\gamma)^{n_1}$	σ_o, K_1, n
Herschel-Bulkley		
Power Series	$\gamma = K_1\sigma + K_2(\sigma)^n$	σ, K_1, n
Carreau	$\sigma^{n_1} = (\sigma_o)^{n_1} + K_1(\gamma^{n_2})$	σ_o, K_1, n_1, n_2
	$\gamma = K_1\sigma + K_2(\sigma)^3 + K_3(\sigma)^5 + ...$ $\sigma = K_1\gamma + K_2(\gamma)^3 + K_3(\gamma)^5 + ...$	$\sigma, \gamma, K_1, K_2, K_3$
	$\mu = \mu_\infty + (\mu_0 - \mu_\infty)[1 + (K_1\gamma)^2]^{\frac{(n-1)}{2}}$	$\mu, \mu_0, \mu_\infty, n, K_1$
Cross	$\mu = \mu_\infty + \frac{(\mu_0 - \mu_\infty)}{1 + (K_1\gamma)^n}$	$\mu, \mu_0, \mu_\infty, n, K_1$
Reiner-Philippoff	$\sigma = \left(\mu_\infty + \frac{(\mu_0 - \mu_\infty)}{1 + ((\sigma)^2/K_1)}\right)\gamma$	$\mu_0, \mu_\infty, \gamma, K_1$

\(Mezger, 2006)

Table 2-7: (continued)

*K_1, K_2, K_3 y n_1, n_2 are arbitrary constants and power indices, respectively, determined from the experimental data.

The second group corresponds to time-dependent fluids. In this type of fluid, the strain rate is not a single-valued function of stress and varies according to the history of previous states. They possess short-term rheological memory, which translates into a measure of apparent viscosity. If the viscosity decreases over time at a given strain rate, it is a thixotropic fluid; if it increases, the fluid is called rheopectic. The behaviour of thixotropic and rheopectic models are unusual or rarely encountered in food products because the phenomena do not govern over time or are not predominant. The behaviour of fluids with this viscosity variation is highly dependent on history and different curves could be obtained for the same sample depending on the experimental procedure (Garriga, 2002).

Due to the importance of rheological properties in the processing of food suspensions, rheological models are built to represent rheological data. Numerous rheological models have been used to describe the flow behaviour in fruit juices and food suspensions such as Newton's law, Herschel-Bulkey, power law, Bingham, Croosy Casson (Fasolin and Cunha, 2012; Vandresen et al., 2009; Quek *et al.,* 2013; Shamsudin *et al.,* 2013). In general, most food fluids have non-Newtonian behaviour. The power law model is the most widely used to describe the rheological behaviour of most fruit juices, especially in handling, heating and cooling operations, as it is convenient, simple and easy to use (Gratáo *et al.,* 2007).

1.6.5 Viscoelasticity

A body under the action of an external force can ideally exhibit two extreme behaviours: elastic and viscous. The elastic behaviour is described by *Hooke*'s law, so that the internal stress is directly proportional to the instantaneous deformation, and is characteristic of pure solids, in which the deformation energy is fully recovered when the force disappears, recovering the original shape. The viscous behaviour is characteristic of pure fluids, which deform in a non-reversible way, since the deformation energy is dissipated as heat, and the original shape is not recovered when the force disappears. These fluids follow *Newton*'s law, such that the internal stress is directly proportional to the deformation gradient, but independent of the deformation itself (Rao, 2006).

In general, materials have an intermediate behaviour between these two extremes, dissipating deformation energy as they flow, while storing energy to partially recover their original shape when the stress disappears. These materials are called *viscoelastic.* The study of viscoelastic properties relates shear stress, strain and time by means of a *rheological equation of state.* To facilitate the study by means of linear differential equations, initially the region of linear viscoelasticity (RVL) is established, a range in which the relationship between stress and strain is only a function of time or frequency, but does not depend on the magnitude of the applied stress (Fig. 2-3) (Steffe, 1996) (Steffe, 1996).

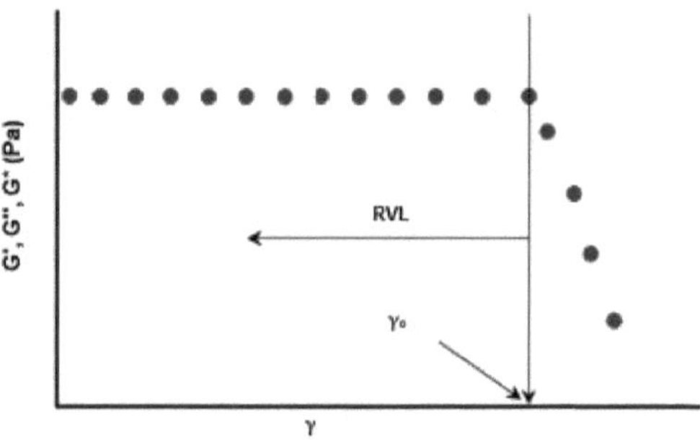

Figure 1-3. Illustration of a typical strain sweep test used to determine the linear viscoelastic limit of a material (Steffe, 1996)

In order to provide information for very short times, the stress or strain can be varied periodically, following a sinusoidal function at a given frequency. A frequency oscillatory experiment is qualitatively equivalent to a time transient (time inverse) experiment. The magnitude of the applied strain or stress must be minimal, in order to remain in the RVL range, i.e. apply a strain following the function (Eq. 2.7):

$$\dot{\gamma}(t) = \dot{\gamma}_0 \, sen(\omega t) \qquad \text{(Ec. 2.7)}$$

Where γ_0 *is* the amplitude and ω is the frequency of oscillation, measuring the effort necessary to maintain this deformation and its variation with frequency. In the case of an elastic solid, the shear stress will be maximum when the deformation is at its maximum, that is, when *sin (ωt)=l*, and therefore, *ωt* is an odd number of times π/2. The response of the material with the applied disturbance will be (Eq. 2.8):

$$\sigma(t) = \sigma_o \, sen(\omega t) \qquad \text{(Ec. 2.8)}$$

If the material is pure viscous, the shear stress will be maximum when the strain rate is maximum, which, being the derivative of the strain with respect to time, is given by the expression (Eq. 2.9):

$$\dot{\gamma} = \frac{d\dot{\gamma}}{dt} = \dot{\gamma}_0 \omega[\cos(\omega t)] \qquad \text{(Ec. 2.9)}$$

That is, it follows the function (Eq. 2.10):

$$\sigma = \sigma_0 \cos(\omega t) = \sigma_0 sen\left(\omega t + \frac{\pi}{2}\right) \qquad \text{(Ec. 2.10)}$$

From these equations it can be seen that in the case of an elastic solid the shear stress is in phase with the deformation, while for a viscous fluid there is a phase difference of π/2 radians. Therefore, a viscoelastic fluid will have a phase shift between 0 and π/2, which will indicate the relationship between elasticity and viscosity, and will depend on the frequency of oscillation. For example, at very high frequencies, corresponding to very short times, the material does not have time to relax and its behaviour is close to that of an elastic solid, with a small phase angle. Conversely, at low frequencies, the material has time to relax and flow and therefore behaves more viscous, which implies a larger phase angle (Fig. 2-4) (Steffe, 1996).

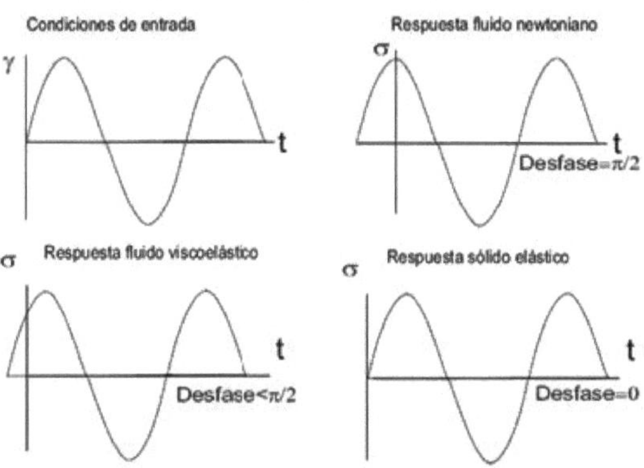

Figure 1-4: Behaviour of a material subjected to an oscillatory test (Rao, 2006).

If δ is the phase angle, which depends on ω and it is taken into account that σ_0 is proportional to γ_0 and also depends on ω, σ can be expressed as (Eq. 2.11):

$$\sigma = G^*(\dot{\gamma}_0)[sen(\omega t + \delta)] \qquad \textbf{(Ec. 2.11)}$$

Where G* is the constant of proportionality between the stress and strain amplitudes, and is called the complex modulus. However, the storage modulus (G') and loss modulus (G") given by the expressions (Eq. 2.12, Eq. 2.13) are more commonly used:

$$G' = G^* \cos(\delta) \qquad \textbf{(Ec. 2.12)}$$

$$G'' = G^* sen(\delta) \qquad \textbf{(Ec. 2.13)}$$

The G' value is a measure of the strain energy stored by the sample during the deformation process. After the load is removed, this energy is fully available, acting now as the driving force for the reforming process that partially or completely compensates the previously obtained deformation of the structure. Materials that store all the energy during deformation exhibit fully reversible deformation behaviour. Therefore, G' represents the elastic behaviour of a material (Mezger, 2006).

The G-value is a measure of the strain energy used by the sample during the cutting process, and is therefore lost to the sample afterwards. This energy is expended during the process of changing the material structure, for example, when the sample is partially or completely flowing. The energy consumed during this friction process is said to be "dissipated". Some of this energy heats the material, and some is lost to the surrounding medium in the form of heat. Therefore, G" represents the viscous behaviour of a material (Mezger, 2006).

Another popular material function used to describe viscoelastic behaviour is the tangent of the phase shift or phase angle (called tan delta), which is also a function of frequency (Eq. 2.14):

$$\tan(\delta) = \frac{G''}{G'} \quad \text{(Ec. 2.14)}$$

If G'> G, elastic behaviour predominates over viscous behaviour. This type of material exhibits a certain stiffness, and is characteristic of stable solids or pastes. However, many dispersions such as pharmaceuticals, lotions or food products show a low to medium viscosity flow tendency and high strain rates, and G'>G" (tan δ<1.0) in the RVL. In this case the material is said to possess a weak gel-like structure, and is conceived as some form of stability (Mezger, 2006). This behaviour has been characteristic in juices, pulps and other fruit-derived suspensions (Meng and Rao, 2005; Moraes *et al.*, 2011; Chaikham and Apichartsrangkoon, 2012; Abbasi and Mohammadi, 2013; Augusto *et al.*, 2013; Moelants *et al.*, 2013; Ma *et al.*, 2013). This same behaviour was found in aloe vera gel in a temperature range of 15 to 45° C. Benchabane and Bekkour (2008) highlight the dominant effect of elastic modulus compared to viscous modulus in low concentrated CMC suspensions (<2.0%). Song *et al.* (2006), in viscoelastic xanthan gum suspensions of less than 3%, cite the importance of the dominance of the elastic modulus with loss tangent values of less than unity (tan δ<0.5).

If G">G' the viscous behaviour dominates the elastic one. The material exhibits the behaviour of a liquid at the RVL (tan δ>1.0). These materials generally do not exhibit stable behaviour at rest, although this behaviour can occur in fluids at a very low flow rate. All fluids that exhibit zero shear viscosity when at rest, i.e., evaluated at a strain rate ≤ 0.01 s^{1} exhibit this behaviour. Other examples include polymer solutions, unbound polymers (silicone), highly viscous varnishes, printing inks, among others (Mezger, 2006).

2 Chapter

EFFECT OF THE INCORPORATION OF HYDROCOLLOIDS AND ALOE VERA ON THE PHYSICOCHEMICAL PROPERTIES AND DEGREE OF STABILITY OF TREE TOMATO *(CYPHOMANDRA BETACEA)* BEVERAGES

ABSTRACT - The effect of the incorporation of hydrocolloids (xanthan gum, CMC) and aloe vera *(Aloe barbadensis Miller)* on the physicochemical characteristics and degree of stability of tree tomato *(Cyphomandra betacea)* beverages was evaluated. High concentrations of hydrocolloids controlled the phase separation of the beverages during storage. Concentrations of xanthan gum and CMC at 0.05% are postulated as suitable treatments in the physical stability of the beverages, expressed in low settling velocities and high Z-potential values (>30mV), without significantly affecting physicochemical properties, colour parameters and sensory attributes. The incorporation of aloe vera gel significantly affected pH and titratable acidity (p>0.05), but its effect on physical instability was not significant. The results suggest that due to the hydrophilic and anionic character of hydrocolloids, they are effective in spherical and electrostatic stability processes, preserving the physicochemical and sensory properties of tree tomato beverages.

Keywords: Hydrocolloids, sedimentation, Z-potential, aloe vera, colour.

ABSTRACT - The effect of the addition of hydrocolloids (xanthan gum, CMC) and aloe vera (Aloe barbadensis Miller) on the physicochemical characteristics and degree of stability of beverages tree tomato (Cyphomandra betacea) was evaluated. High concentrations of hydrocolloids controlled phase separation of beverages during storage. Concentrations of xanthan gum and CMC 0.05%, are suggested as adequate physical stability in the beverage treatments, expressed at low settling rates and high values of zeta potential (> 30mV), without significantly affecting physicochemical properties, parameters colour and sensory attributes. The addition of aloe vera gel significantly affect the pH and titratable acidity (p> 0.05), but its effect on the physical instability was not significant. The results suggest that due to the hydrophilic character and anionic hydrocolloids, are effective in the process of electrostatic and steric stability, whilst maintaining the physicochemical and sensory properties of tree tomato beverages.

Keywords: Hydrocolloids, sedimentation, zeta potential, aloe vera, colour.

INTRODUCTION

In Colombia, the tamarillo or tree tomato *(Solanum betaceum cv. sendtn)* is a promising product for export, and due to its colour, the orange variety has been the most internationally accepted. Its flesh is highly appreciated for its pleasant aroma, bittersweet-astringent taste and orange colour (Meza and Manzano, 2009). Tree tomatoes are known for their high content of vitamins, minerals, carotenoids, anthocyanins and other phenolic compounds (Márquez *et al.*, 2007; Nascimento *et al.*, 2013; Castro *et al.*, 2013). These compounds are not only responsible for the colour of the fruit, but also possess biological, therapeutic and preventive properties (Castro *et al.,* 2013). The pulp is used in the production of juices, nectars, jams, salads and preserves (Meza and Manzano, 2009).

Fruit pulp has a large amount of insoluble polymeric material, which increases physical instability leading to phase separation during storage, particularly in food suspensions (Genovese *et al.,* 1997). Several studies highlight the functional properties of aloe vera, and mention that the inclusion of such components in food strengthens its nutritional value (Habeeb *et al.,* 2007; Ramachandra *et al.,* 2008; Boghani *et al.,* 2012). However, aloe vera gel (*Aloe barbadensis* Miller) is a mucilaginous product containing polymeric materials, amino acids, lipids, minerals and vitamins, which can affect the stability of dispersions (Yaron *et al.,* 1992).

Phase separation in liquid product development is a quality defect, affecting product marketability and acceptability (Ibrahim *etal.,* 2011). Since the particles in fruit nectars are negatively charged, the addition of anionic hydrocolloids is expected to increase the electrostatic repulsion forces between particles. Xanthan gum and Carboxymethylcellulose

(CMC) are hydrophilic and negatively charged polysaccharides, leading to spherical and electrostatic stabilisation of insoluble particles (Genovese and Lozano, 2001; Liang *et al.*, 2006). Likewise, hydrocolloids, due to their high water-holding capacity, allow regulating the rheological and textural characteristics of food systems, increasing viscosity, creating a gel-like structure and providing physical stability (Sahin and Ozdemir, 2007). There is a large body of research related to the use of hydrocolloids (such as xanthan gum, CMC, guar gum, gellan, pectin, chitosan, among others) in food dispersions (Genovese and Lozano, 2001; Liang *et al.*, 2006; Ibrahim *et al.*, 2011; Sahin and Ozdemir, 2007; Ghafoor *et al.*, 2008; Nwaokoro and Akanbi, 2015; Abbasi and Mohammadi, 2013; Lins *et al.*, 2014).
The aim of this study was to evaluate the effect of the addition of hydrocolloids and aloe vera on the physicochemical properties of tree tomato beverages, and their impact on the degree of stability of tree tomato beverages.

MATERIALS AND METHODS

Tree tomato *(Cyphomandra betacea)* fruits of the common orange variety were used. The fruits were selected with a maturity grade, colour typology 6 (NTC 4105/1997). Hydrocolloids (xanthan gum, sodium carboxymethylcellulose-CMC) and food grade preservatives (sodium benzoate and potassium sorbate) supplied by Bell Chem International S.A. Food grade Aloe vera gel (*Aloe barbadensis Miller)* with 98% purity, was purchased from Ayala Benard S.A.S.

Characterisation of raw materials: The physicochemical characterisation of tree tomato pulp and aloe gel was carried out by evaluating properties such as pH, total soluble solids (TSS) and titratable acidity using AOAC (2005) protocols. Viscosity was also calculated by rheometry and yield, expressed as a percentage (Márquez, 2009).

Preparation of the nectar drink: The plant material was disinfected in sodium hypochlorite solutions at 100 ppm, for 10 minutes, and finally washed in water. The pulp was extracted in a pulping machine (D1000, CITALSA, Colombia). The beverages were formulated from a 300 g base, with a pulp content of 18% (NTC 3549/1999). Sucrose was used as sweetener, and water was added as solvent until a beverage with a soluble solids concentration of 10° Brix was obtained. The respective amount of hydrocolloids was added, after preparation in a water suspension at 40° C. In addition, a preservative mixture equal to 0.125% was added to prolong the shelf life and to study the degree of stability under storage conditions. The beverages were homogenised in a ULTRA-TURRAX disperser (IKA, T25 Basic, Germany) for 60 s at 5000 rpm. They were also subjected to a mild pasteurisation process at 60°C for one minute. The final product was packed in plastic containers (PET) and stored under refrigeration for characterisation.

Physicochemical properties: The tree tomato beverages were physicochemically characterised, determining the following properties based on la AOAC (2005) standard: Soluble solids concentration (TSS) expressed as° Brix; hydrogen potential (pH); titratable acidity expressed as percentage of citric acid; and density by pycnometry using distilled water as reference liquid. Measurements were performed in triplicate, at room temperature (25°C).

Zeta potential determination: Zeta potential was determined by electrophoresis using the Zeta-sizer instrument (Model ZS90, Malvern, UK) following the methodology proposed by Genovese and Lozano (2001). Samples were diluted in distilled water at a ratio of 1:20. The study was performed in triplicate at room temperature (25°C).

Particle size: The particle size of the beverages (PSD) was determined by light scattering using a Mastersizer (Model 3000E, Malvern, UK) guided by the methodology proposed by Kaneiwa *et al.* (2013). Particle size will be expressed as a function of mean diameter based on surface area ratio (D [3.2]) and volume ratio (D [4.3]) according to the expressions (Eq. 3.1, Eq. 3.2), where n_i is the number of particles with diameter d_i .

$$D[3,2] = \frac{\sum_i n_i d_i^3}{\sum_i n_i d_i^2} \quad \text{(Ec. 3.1)}$$

$$D[4,3] = \frac{\sum_i n_i d_i^4}{\sum_i n_i d_i^3} \quad \text{(Ec. 3.2)}$$

Both ratios were studied, since D [4,3] is influenced by large particles, while D [3,2] is more influenced by small particles (Bengtsson *etal.,* 2011).

Sedimentation test: Phase separation was established based on the change in height observed in 50 mL graduated cylinders, based on the protocol described by Fasolin and Cunha (2012). The analyses were carried out at a temperature of 10 C±2^{oo} C, for 120 hours until phase equilibrium was reached. Phase separation was observed visually, and the sedimentation index *(SI)* was calculated using the following expression (Eq. 3.3):

$$IS(\%) = \left(\frac{H_t}{H_o}\right) * 100 \quad \text{(Ec. 3.3)}$$

Where H_t is the height of the upper phase at the interface after a time ty H_0 is the initial height. The influence of the compositions of the formulations on the sedimentation process was analysed by means of a first order kinetic equation (Eq. 3.4).

$$IS = IS_{eq}(1 - e^{-vt}) \quad \text{(Ec. 3.4)}$$

Where *IS* is the volume of sediment at equilibrium, *t* the time, and *v* the sedimentation velocity in h^[1] .

Instrumental colour assessment: Colour measurement was performed by the CIELAB tristimulus system (L*, a*, b*) using a sphere spectrophotometer (Model CR400, Konica-Minolta) guided by the methodology proposed by Fernández *et al.* The results were expressed in conformity with reference to the D65 illuminant and with a visual angle of 10°. The measurements were performed in triplicate. From the uniform CIELAB colour space, the colour difference (ΔE) defined by the following expression (Eq. 3.5) was determined.

$$\Delta E = \sqrt{(\Delta L^*)^2 + (\Delta a^*)^2 + (\Delta b^*)^2} \quad \text{(Ec. 3.5)}$$

Turbidity (Cloudiness): The beverages were sedimented in an accelerated manner using a centrifuge (Hettich Universal, Model 320R, Germany) at 4200 *g* for 15 min in 25 ml graduated tubes. The supernatant was collected and the absorbance was determined at 660 nm in a UV-visible spectrophotometer (Thermo Scientific, Evolution 60S, USA), using distilled water as a standard. The result was directly related to the turbidity or cloudiness of the supernatant, since suspended particles absorb more radiation, and is inversely proportional to sedimentation (Silva *et al.,* 2010; Kubo *et al.,* 2013).

Phase characterisation: The phases (sediment and supernatant) obtained after the centrifugation sedimentation process, were characterised through the evaluation of parameters such as density, Z-potential and particle size distribution, as described in the sections described above. Viscosity was determined using a viscometer (Brookfield, Model DV III Ultra, USA) with concentric cylinder geometry, at 25° C and a shear rate of $10s^{-1}$.

Sensory analysis: A sensory evaluation was carried out using a descriptive, parametric test with an unstructured scale (Márquez, 2009). Parameters such as colour, flavour, aroma and appearance were evaluated. The evaluation was carried out with 15 semi-trained judges selected for their knowledge of the sensory characteristics of tree tomato pulp. After the judge's weighting per attribute, the scores were studied and the respective statistical analysis was applied.

Experimental design: A central rotational design with axial points was established for the experimentation, with four replicates at the central point. The factors and levels established in the study are defined in Table 3-1. The results were statistically analysed by generating response surfaces, analysis of variance, lack of fit test and determination of regression coefficients using Statgraphics Centurion XVI software, version 16.1.18.

Table 2-1: Factors analysed and coding of the established levels

Factor	Symbol	Under	Medium	High
Coding		-1	0	+1
Aloe vera (%)	X_i	0,5	1,0	1,5
Xanthan gum (%)	X_2	0,025	0,05	0,075
Sodium carboxymethyl cellulose - CMC (%)	X_3	0,025	0,05	0,075

The experimental data were fitted to the following second-order model (Eq. 3.6):

$$Y = \beta_0 + \sum_{i=1}^{3}\beta_i X_i + \sum_{i=1}^{3}\beta_{ii} X^2{}_i + \sum\sum_{i<j=1}^{3}\beta_{ij} X_i X_j \qquad \text{(Ec. 3.6)}$$

Where $\beta_o, \beta_i, \beta_{ii}, \beta_{ij}$ are the regression coefficients for the intercept, linear and quadratic interaction terms, respectively, andX is the independent variable.

ANALYSIS AND RESULTS

Characterisation of raw materials: Table 3-2 details some physicochemical characteristics of tree tomato pulp and aloe vera gel. The values of pH, titratable acidity and solids concentration are close to those reported by Márquez *et al.* (2007) for tree tomato and Elbandy *et al.* (2014) for aloe vera juice. However, it should be noted that these characteristics may differ as they are a function of variety, degree of maturity or ecophysiological growing conditions.

Table 2-2: Physicochemical properties of tree tomato *(Cyphomandra betacea cv. Sendtn)* pulp and aloe vera *(Aloe barbadensis Miller)* gel.

Feature	Pulp	Gel
P^H - .	3,44 ± 0,39	4,33 ± 0,53
SST (° Brix)	11,20 ±1,10	4,20 ± 0,32
Titratable acidity (% citric acid)	1,66 ± 0,08	1,12±0,02
Viscosity (mPa.s)	2634 ± 87	3,10 ± 0,27
Yield (%)	69,35±2,62	-

Arithmetic mean ± standard error.

Physicochemical properties: The results of physicochemical characteristics of tree tomato beverages formulated with hydrocolloids and aloe vera are detailed in Table 3-3. The addition of hydrocolloids and aloe vera did not exert a significant effect ($p>0.05$) on the behaviour of the physicochemical properties of the beverages. Similar results were found by Chatterjee *et al* (2004). They report that in clarified fruit juices, TSS concentration and titratable acidity percentage did not change significantly after the addition of chitosan ($p>0.05$). Chaikham and Apichartsrangkoon (2012) report that xanthan gum did not influence the pH and titratable acidity titration of *Dimocarpus longan* juices. At the same time, it is reported that the contents of acidity, TSS and pH value show a defined behaviour, independently of *Enterolobium cyclocarpum* gum concentration, in peach juices (Delmonte *etal.,* 2006).

Table 2-3: Physico-chemical properties of tree tomato beverages

Essay	Xi	X_2	X_3	pH	Titratable acidity (% citric acid)	Density (g/mL)	SST (° Brix)	Potential Z "
T1	0	0	1,68	4,28	0,49	1,031	9,57	-36,90
T2	0	-1,68	0	4,31	0,51	1,029	9,37	-33,20
T3	-1,68	0	0	4,35	0,53	1,025	9,80	-37,40
T4	-1	1	1	4,34	0,51	1,026	9,83	-35,93
T5	1	-1	1	4,25	0,52	1,009	9,47	-29,33
T6	0	0	0	4,19	0,49	0,991	9,33	-38,27
T7	1	1	-1	4,11	0,52	1,021	9,40	-38,90
T8	0	0	0	4,10	0,49	0,988	9,33	-36,46

T9	1	-1	-1	4,19	0,52	1,007	9,37	-30,53
T10	0	0	-1,68	4,19	0,50	1,024	9,67	-34,23
T11	0	1,68	0	4,31	0,52	1,022	9,70	-41,00
T12	1,68	0	0	4,31	0,54	1,027	9,70	-36,20
T13	0	0	0	4,09	0,51	1,021	9,53	-34,93
T14	0	0	0	4,09	0,51	0,977	9,40	-35,12
T15	-1	-1	1	4,26	0,44	1,029	9,60	-36,23
T16	1	1	1	4,21	0,53	1,024	9,63	-42,67
T17	-1	-1	-1	4,32	0,54	0,977	9,57	-27,57
T18	-1	1	-1	4,25	0,54	1,033	9,43	-39,80
Control	-	-	-	4,35	0,49	1,047	9,90	-18,17

However, the concentration of aloe vera proved to be significant in the variation of pH and titratable acidity of the formulated beverages ($p<0.05$), a result that is evident in Fig. 3-1, where a significant decrease in pH related to an increase in the percentage of acidity is observed. This is probably due to the presence of organic acids constituents of the aloe vera gel, intensifying the acidity of the formulated beverages. Bozzi *et al.*, (2007) point out that aloe vera gel contains citric acid, acetic acid and malic acid, which may be responsible for the decrease in pH. In addition, a concentration of 1.23% citric acid has been reported in aloe juice (Boghani, 2012).

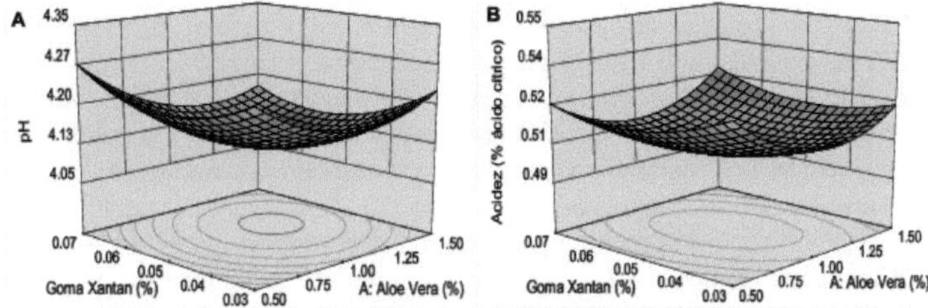

Figure 2-1: Response surfaces of physicochemical property behaviour of tree tomato beverages
at a CMC concentration of 0.50%, for: (A) pH; (B) Acidity.

The pH value of the beverages fluctuated between 4.01 and 4.31, values lower than those estimated for the control beverage. Likewise, there was an increase in titratable acidity in the treated beverages compared to the control, differences associated with the effect of the properties of the aloe gel. Elbandy *et al.* (2014) reported a significant increase in titratable acidity and pH titration in mango nectars formulated with aloe vera. In turn, Hamid *et al.* (2014) report an increase in acidity in mixed carrot-orange nectars, which they attribute to the content of acids (salicylic, uranic, phenolic) present in the aloe gel. The coefficients of determination (R^2) of the fit model for pH and titratable acidity behaviour, respectively, were greater than 0.75 (Table 3-4).

Table 2-4: Regression coefficients for physicochemical properties performance of tree tomato beverages.

Regression coefficients	pH	Acidity	Potential Z
βo	4,83	+0,63	-25,33
ßl	-0,55	-0,15	3,10
ß2	-9,80	-1,26	-111,73
β_3	-6,68	-0,63	-216,26
ßn	-1,31	0,22	-97,68
β_{l2}	1,21	0,54	22,32
β_{13}	40,20	-2,00	NS
β_{22}	0,25	0,06	NS

β_{23}	87,43	11,12	-31,82
β_{33}	44,43		NS835,66
R^2	0,80		0,860,75
Model (p-value)	0,04	0,07	0,04
Missing fit (p-value)	0,36	0,08	0,11

NS: non-significant coefficients

On the other hand, the concentration of aloe vera did not exert a significant effect on the TSS and density behaviour of the beverages. Elbandy *et al.* (2014) report that the addition of aloe up to a concentration of 10% did not influence the behaviour of TSS content in mango nectars. They state that aloe vera gel has a low solids content (<4.0%), which when incorporated into food matrices slightly affects the physicochemical behaviour of the product. In turn, Subhra *et al.* (2014) found no differences in TSS variation in mandarin (Kinnow) nectars enriched with 4% aloe gel. The coefficients of determination of the fit model for TSS concentration and density in tree tomato beverages were greater than 0.80 (Table 3-4). The fit test allows us to infer that the estimated models adequately describe the trend of the physicochemical variables as a function of the addition of hydrocolloids and aloe vera ($p>0.05$).

Zeta potential: Z-potential measurements are used to evaluate the stability of colloidal systems, since it is a good index of the magnitude of the interaction between particles in the dispersion. From the statistical analysis, the significant effect of xanthan gum ($p<0.05$) on the decrease of zeta potential and stability of the suspensions is highlighted (Fig. 3-2). This is probably due to the fact that xanthan gum, being an anionic (or negatively charged) polysaccharide, has the ability to induce highly repulsive forces preventing colloidal particles from binding, controlling certain mechanisms of stability breakdown in suspensions (Mirhosseini and Tan, 2010). Genovese and Lozano (2001), highlight the significant effect on colloidal stability of apple juices formulated with 0.4% xanthan gum. Hydrocolloids can stabilise dispersions through viscous effects, spherical hindrance and electrostatic interactions (Dluzewska *et al.*, 2005). In this case, given the increase in Z-potential values due to the incorporation of xanthan gum, we can infer that there are dominant mechanisms associated with electrostatics.

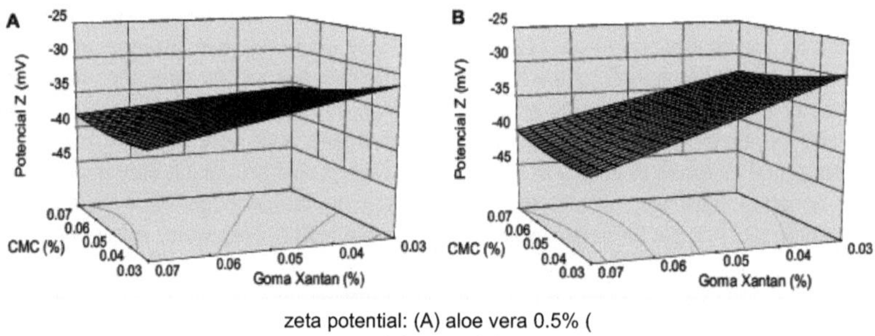

zeta potential: (A) aloe vera 0.5% (
minimum point
), (B) aloe vera 1.5% (maximum point)

Table 3-4 shows the regression coefficients of the mathematical model that describes the behaviour of Z-potential as a function of the use of hydrocolloids and aloe vera. A coefficient of determination (R^2) of 0.76 was estimated and the test of fit allows inferring that the selected model is adequate to describe its behaviour ($p>0.05$). The Z potential values in the tree tomato beverages ranged from -27.57 to -42.67 mV. These results are close to those reported for apple juices stabilised with xanthan gum and CMC (Genovese and Lozano,

2001). Also, Z-potential values between -24.6 and -39.1 are reported for orange drinks formulated with pectin and CMC (Dluzewska *et al.,* 2005). Leiberman *et al.,* (1998) argue that absolute values greater than 25mV in the Z-potential measurement are indicative of stable emulsion systems with no clumping effect.
systems with no agglomeration effect. Abbasi and Mohammadi (2013) highlight the stabilising power of persian gum in milk and orange juice emulsions from Z-potential values that ranged from -24 to -31 mV. Consequently, the stabilising capacity of hydrocolloids such as xanthan gum and CMC in beverages can be highlighted, with respect to the control which reached a value equal to -18.16 mV. Schramm (2005), states that in suspensions with Z-potentials lower than -20mV, instability mechanisms such as agglomeration and precipitation of particles predominate.

Particle size: In table 3-5, the particle sizes for the different formulated nectars are found. It is evident that the volume-based mean diameter D [4,3] (influenced by larger particles) showed higher values than the surface-based mean diameter D [3,2] (influenced by smaller particles), which explains the size effect in each estimation method.

Table 2-5: Particle size and colour parameters in tree tomato beverages

Essay	Xi	X_2	X_3	D [3,2], µm	D [4,3], µm	L*	a*		b* ΔE* ΔE* ΔE* ΔE*
T1	0	0	1,68	143	250	39,04	4,50	12,46	0,89
T2	0	-1,68	0	145	243	39,12	4,12	11,06	0,92
T3	-1,68	0	0	145	265	39,00	3,73	10,63	0,87
T4	-1	1	1	148	246	39,03	3,89	10,53	0,86
T5	1	-1	1	142	240	40,84	4,12	10,83	2,44
T6	0	0	0	142	237	38,27	4,18	10,64	0,56
T7	1	1	-1	147	241	39,76	3,31	9,92	1,68
T8	0	0	0	140	243	38,19	4,87	10,62	1,15
T9	1	-1	-1	142	247	38,86	4,06	10,68	1,65
T10	0	0	-1,68	139	248	40,28	3,98	12,81	1,26
T11	0	1,68	0	150	260	38,96	3,56	10,09	1,12
T12	1,68	0	0	145	241	38,83	3,58	10,16	0,80
T13	0	0	0	144	235	39,67	3,82	9,98	1,46
T14	0	0	0	145	247	36,47	4,19	10,69	0,87
T15	-1	-1	1	146	250	41,99	3,21	10,02	1,22
T16	1	1	1	148	242	42,85	3,02	9,80	2,10
T17	-1	-1	-1	136	256	43,66	3,97	12,48	2,24
T18	-1	1	-1	140	256	40,55	3,08	10,08	1,72
Control	-	-	-	139	231	36,58	4,79	12,56	1,47

The addition of hydrocolloids and aloe vera did not exert a significant effect on the variation of the mean diameters of the suspensions ($p>0.05$). Similar results were reported by Sherafati *et al.* (2013) on carrot juice stabilised with gellan gum. Particle size showed a monomodal distribution between ~80 and 800 µm, similar for all beverages (Fig. 3-3). This can be explained by the standard homogenisation process applied to the different beverages, contributing to a better uniformity of the particle size distribution. Silva *et al.,* (2010) state that homogenisation is a mechanical process involving the reduction of particles or droplets to create a stable dispersion.

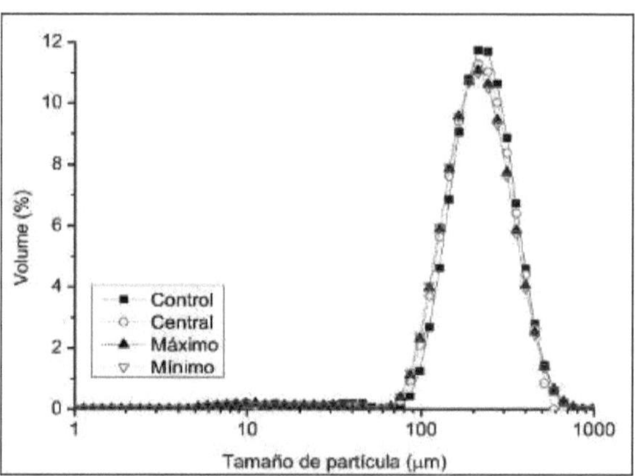

Figure 2-3: Particle size distribution in tree tomato beverages

The mean diameters based on surface area D [3,2] and volume D [4,3], in the beverages ranged from 136-150 µm and 235-265 µm, respectively. These values are close to those reported by Kaneiwa *et al.,* (2013) in non-homogenised tomato juices at high pressures. Leite *et al.,* (2014) estimated mean values of diameter D [4,3] between 228-278 µm in orange juices. However, the mean diameter D [3.2] in these same beverages are relatively lower than those found in this study. When comparing the treated beverages with the control, an increase in the mean diameters is evident (Table 3-5). Huang *et al.* (2001) found that the incorporation of hydrocolloids increased the particle size in emulsions. Keshtkaran *et al.,* (2013) argue that the incorporation of polymeric materials into suspensions can produce variation in particle size distribution in suspension, due to insoluble fractions of polysaccharides and cellulosic substances.

Stoke's law states that the sedimentation rate is proportional to the size of particles in suspension. In turn, Schramm, (2005) states that dispersions with a distribution towards smaller sizes will represent greater stability. However, since hydrocolloids tend to increase mean particle diameters, they can give the dispersion stability during storage (Huang *et al.,* 2001; Homayoonfal and Mousavi, 2015). Koyama and Kitamura (2014), were able to stabilise rice beverages with particle diameters ranging from 1.0 to 100 µm, with xanthan gum concentrations above 0.1%.

Sedimentation rate: Table 3-6 describes the kinetic parameters of the simulation model describing the phase separation phenomenon in tree tomato beverages during storage, as a function of the sedimentation rate reached equilibrium (SI_{eq}) and sedimentation velocity (v). The sedimentation rate (SI) decreased with hydrocolloid concentration (Fig. 3-4). Nwaokoro and Akanbi (2015) report control in sedimentation level in mixed tomato-carrot juice stabilised with CMC, xanthan and guar. Likewise, Ghafoor *et al.* (2008) highlight the stabilising power of xanthan gum, gellan and carrageenan on the sedimentation of grape juice. On the other hand, it is worth noting that some nectars did not show sedimentation (IS=O) during the period evaluated. These results were found for high concentrations of xanthan gum and CMC. This is probably due to the synergistic effect of the stabilisers, since xanthan gum has the capacity to increase the apparent viscosity of the dispersed phase, while CMC, being an electronegative hydrocolloid when added to suspensions, increases the electrostatic repulsion forces between particles, giving stability to food suspensions (Genovese and Lozano, 2001).

Table 2-6: Sedimentation rate and sedimentation velocity (v) in tree tomato beverages.

Essay	Xi	X_2	X_3	IS	$v(h^{-1})$*	Turbidity (Abs.)
T1	0	0	1,68	0,0	0,00	0,954
T2	0	-1,68	0	40,9	0,127	0,195
T3	-1,68	0	0	4,5	0,027	0,885
T4	-1	1	1	0,0	0,00	0,762
T5	1	-1	1	37,1	0,103	0,249
T6	0	0	0	4,06	0,024	1,092
T7	1	1	-1	0,0	0,00	0,656
T8	0	0	0	5,0	0,022	0,982
T9	1	-1	-1	39,5	0,113	0,194
T10	0	0	-1,68	7,38	0,064	0,415
T11	0	1,68	0	0,0	0,00	0,997
T12	1,68	0	0	24,5	0,028	0,895
T13	0	0	0	6,4	0,039	0,927
T14	0	0	0	3,50	0,023	0,788
T15	-1	-1	1	22,3	0,093	0,586
T16	1	1	1	0,0	0,00	0,914
T17	-1	-1	-1	55,7	0,130	0,260
T18	-1	1	-1	0,0	0,00	0,894
Control	-	-	-	67,5	1,541	1,406

* Regression coefficients were greater than 0.90.

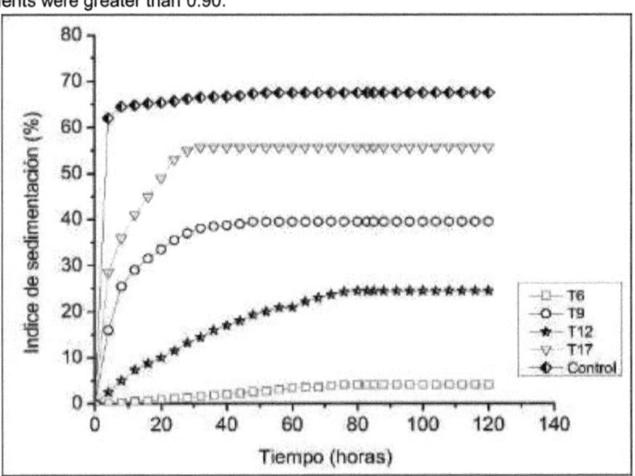

Figure 2-4: Sedimentation rate behaviour of tree tomato beverages during storage at 10 C.°

Table 3-6 details the behaviour of the sedimentation rate for the various treatments carried out. The addition of xanthan gum was found to be significant ($p<0.05$) in reducing phase separation and improving the physical stability of the beverages. Stokes' law indicates that sedimentation of particles in suspensions is inversely proportional to their viscosity, which means that higher viscosity can beneficially increase juice stabilisation (Genovese and Lozano, 2001). Several studies have reported the use of xanthan gum and its ability to improve rheological properties and increase viscosity of dispersions offering better stability of food products (Moraes *et al.,* 2011; Chaikham and Apichartsrangkoon, 2012; Paquet *et al.,* 2014; Cho and Yoo, 2015). In addition, Sahin and Ozdemir (2007) found that the use of xanthan gum is an excellent stabiliser of tomato sauces. They cite that xanthan gum possesses excellent water holding capacity (WRC), a characteristic that makes it effective in preventing phase separation.

In Fig 3-5, a significant decrease in settling velocity is observed with increasing xanthan gum, a result consistent with the increase in Z-potential, indicating an improvement in the stability of tree tomato beverages. Ghafoor *et al.,* (2008) argue that xanthan gum is an anionic polysaccharide, which has the ability to generate electrostatic repulsion between particles, moderating phase separation. Also, given its high molecular weight, it causes an increase in apparent viscosity and sedimentation control in food suspensions (Sahin and Ozdemir (2007). On the other hand, the model describing the behaviour of the settling velocity as a function of hydrocolloids and aloe vera had a coefficient of determination of 0.936 (Table 3-7), and the test of fit was not significant ($p>0.05$). The settling velocity ranged from 0.0 - 0.13 $h^{\wedge 1}$, these values are higher compared to those estimated for soursop juice stabilised with soy protein (Fasolin and Cunha, 2012).

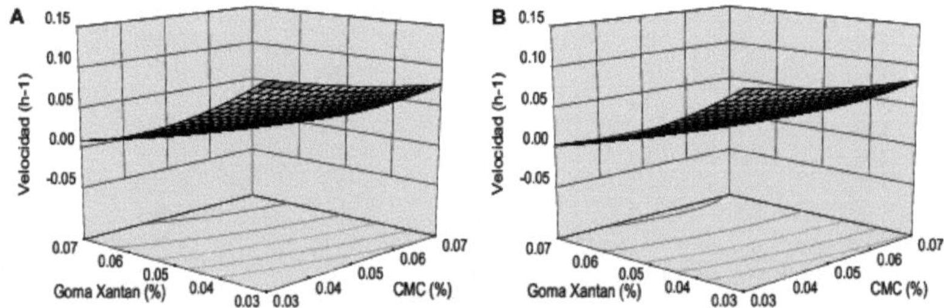

Figure 2-5: Response surfaces for settling velocity behaviour in tree tomato beverages
: (A) Aloe vera 0.5%; (B) Aloe vera 1.5%.

Turbidity (Cloudiness): Table 3-6 details the behaviour of turbidity in the different formulated beverages. The incorporation of xanthan gum was found to be significant ($p<0.05$). Turbidity or cloudiness in suspensions is the result of dispersed insoluble particles (Silva *et al.,* 2010). The incorporation of hydrocolloids produces higher turbidity (absorbance) in the supernatant, indicating a higher retention of suspended particles and a decrease in the phase separation phenomenon. Similar behaviour was reported by Ghafoor *et al.* (2008) in grape juice stabilised with xanthan gum, gellan and carrageenan. The regression coefficients and parameters are shown in Table 3-7. From the test of fit ($p>0.05$), it can be inferred that the regression coefficients satisfactorily describe the derived model.

Table 2-7: Regression coefficients for settling velocity behaviour, particle size and colour parameters.

Regression coefficients	$v(h)^{1}$	Turbidity (Abs.)	Viscosity (mPa.s)
βo	0,28	-0,76	10,41
βl	-0,04	-0,41	-3,22
$\beta 2$	-4,99	31,22	168,89
β_3	-1,93	36,19	484,0
$\beta 11$	0,07	6,79	NS
$\beta l2$	NS	NS	54,80
β_{13}	9,48	-123,4	2696,0

β_{22}	0,01	-0,19	-1,59
$\beta\,3_2$	25,31	NS	3887,96
$\beta\,3_3$	7,29	-368,12	NS
R^2	0,94	0,87	0,98
Model (p-value)	0,00	0,04	0,00
Missing fit (p-value)	0,07	0,29	0,08

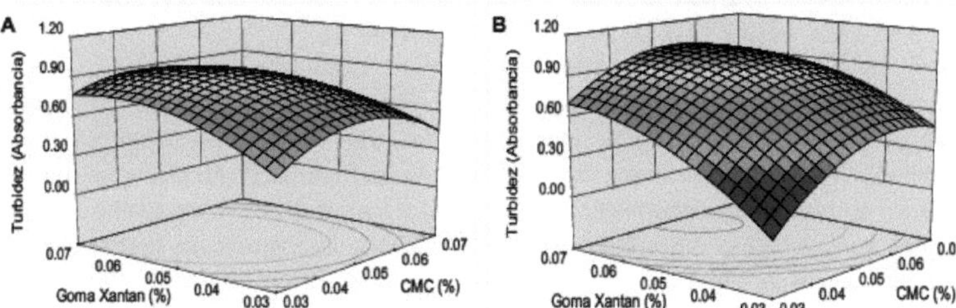

Figure 2-6: Response surfaces for turbidity behaviour in tree tomato beverages: A) Aloe vera 0.5%, B) Aloe vera 1.5%.

Fig. 3-6 shows a significant increase in turbidity as the hydrocolloid concentration gradually increases. Mirhosseini *et al.* (2008) reports an increase in the turbidity of orange emulsions stabilised with pectin and CMC. In turn, Genovese and Lozano (2001) report an improvement in the stability of apple juice by incorporating CMC and xanthan gum, attributed to the increase in turbidity in the clarified phase (supernatant). Authors suggest that xanthan gum and CMC are negatively charged ionic polysaccharides and produce an increase in electrostatic repulsion forces between particles, causing a stability effect in suspensions (Genovese and Lozano, 2001; Ghafoor *et al.,* 2008).

After the mechanical separation process by centrifugation, the tree tomato beverages were able to settle. However, a direct relationship is observed between the values of Z-potential, turbidity and sedimentation rate. Drinks formulated with concentrations higher than 0.05% hydrocolloids showed higher Z-potential values, higher turbidity of the clarified phase and lower sedimentation rates; factors that indicate good stability of suspended particles. Some authors manage to establish similar relationships during the study of stability of fruit juices stabilised with hydrophilic colloids (Genovese and Lozano, 2001; Benítez and Lozano, 2007; Mirhosseini and Tan, 2010).

Phase characterisation: The characterisation of the phases in tree tomato beverages was carried out to better understand the instability phenomenon. There are no significant differences in the behaviour of density, Z potential differential and mean particle diameters in relation to the concentration of hydrocolloids and aloe vera ($p>0.05$). It is observed that both density and mean diameters of the sediment phase are higher than those estimated for the supernatant phase (Table 3-8). These differences can be considered as dominant factors in the occurrence of instability mechanisms in the beverages. A reduction in particle size leads to postponed gravitational separation, because according to Stokes law, the settling velocity is proportional to the square of the particle radius. In addition, smaller particles will have lower weight (lower density), and will precipitate less. McClement (2005), states that the tendency of large particles to coalesce is greater than in small particles, therefore, reducing the size can increase the stability and extend the shelf life of emulsifying beverages. On the other hand, the mean diameters in beverages formulated with hydrocolloids with respect to the control did not show significant differences, possibly associated with the homogenisation process.

Table 2-8: Physicochemical characteristics of supernatant and settled phases in tree tomato beverages.

	Control Drink		Treated beverages	
	Supernatant	**Sediment**	**Supernatant**	**Sediment**
Density (g/ml)	0,99 ± 0,01	1,49 ± 0,12	1,08±0,02*	1,49±0,07*
Potential Z (mV)	-19,13 ± 1,83	-31,37±2,21	-34,14±2,18*	-42,92 ± 3,68*
D [3,2], µm	7,08 ±1,43	151 ± 3,53	9,08 ± 1,44*	157,61 ±4,12*
D [4,3], µm	29,60±2,81	249,56±12,32	35,88±3,15*	271 ± 12,84*
Viscosity (mPa.s)	4,50±0,37	3229,8±172	-	2722,60±110*

Arithmetic mean ± standard error. * Average values of the 18 tests carried out.

The viscosity of the supernatant phase was significant as a function of the addition of xanthan gum and CMC (p<0.05). An increase in viscosity was observed with the addition of hydrocolloids (Fig. 3-7), as these have a high water retention capacity, giving greater consistency to aqueous suspensions. In turn, the increase in viscosity in the supernatant phase may correspond to a greater presence of particles in suspension, higher turbidity, an indicator of stabilising effects in beverages. The coefficients and regression parameters for the behaviour of the viscosity of the supernatant phase are shown in Table 3-7, with a non-significant test of fit (p>0.05). In relation to the Z potential differential in the supernatant phase, a significant increase is observed in the treated beverages with respect to the control, associated with the presence of anionic substances. Consequently, it can be inferred that the physical stability of the beverages with the addition of hydrocolloids is associated with spherical processes (increase in viscosity of the continuous phase) and electrostatic processes (increase in repulsive forces between particles).

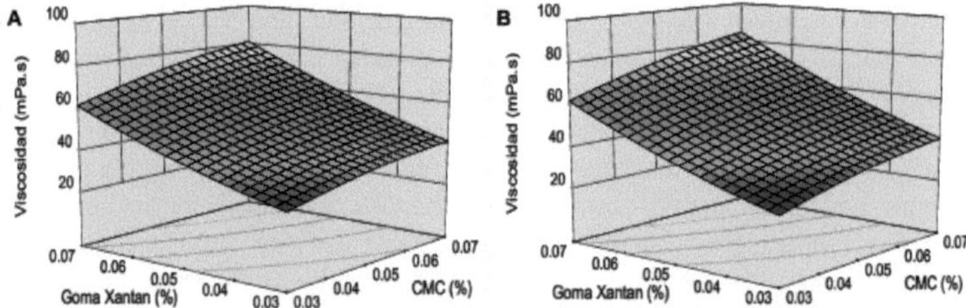

Figure 2-7: Response surfaces for the viscosity behaviour of the supernatant phase in tree tomato beverages: A) Aloe vera 0.5%, B) Aloe vera 1.5%.

The viscosity of the sedimented phase did not show significant behaviour as a function of hydrocolloid and aloe vera concentration (p>0.05). However, the viscosity was lower than that estimated in the control sample, a trend that may be associated with the reduction of phase separation processes or particle sedimentation. In turn, the sedimented phase in the treated beverages shows a higher Z potential differential, possibly due to the presence of anionic hydrocolloids.

Colour determination: Table 3-9 shows the behaviour of the colour parameters in CIELab coordinates, as well as the colour differential (ΔE), of tree tomato beverages. No significant effect of hydrocolloids and aloe vera (p>0.05) on the variation of L*, a*, b* and ΔE parameters was found. However, the lightness (L) values in treated beverages were higher than those estimated in the control beverage. A similar behaviour was reported in chitosan-stabilised enriched orange juices (Martin *et al.,* 2009). The decrease in brightness during storage time in juices is attributed to particle settling during storage time (Genovese *et al.,* 1997). However, L values in treated tree tomato beverages did not show this trend. These results could be explained by the increase in suspension clarity, an effect associated with the use of hydrocolloids (Chatterjee *et al.,* 2004). The increase in L values may be due to the natural colour of biopolymers, which produces a clarity effect in beverages, especially at high concentrations (Martin *et al.*, 2009; Rungsardthong *et al.*, 2006; Halim *et al.*, 2014).

Table 2-9: Colour parameters and sensory evaluation of tree tomato beverages

Parameter	Type of beverage	
	Control	Treaty
L*	36,58±1,06	39,74±1,54**
a*	4,79±0,81	3,85 ± 0,67**
b*	12,56±1,20	10,75±0,73**
ΔE	1,47±0,06	1,32 ± 0,05**
Colour	12,82 ± 1,33	11,49±1,08**
Taste	13,54±1,78	9,54 ± 1,21**
Aroma	12,65±2,33	11,53±1,40**
Appearance	9,34 ±1,54	12,49 ± 1,35**

Arithmetic mean ± standard error. ** Average values of the 18 tests carried out.

Positive a* values ranged from 3.02 to 4.87, indicating a visual position towards red, while positive b* values ranged from 9.80 to 12.81 indicating a tendency towards yellow. A non-significant decrease in a* and b* parameters was observed in treated beverages, compared to the control. Qin *et al.*, (2002), argue that these changes can be attributed to pigment degradation or precipitation. In turn, Halim *et al.* (2014) report a decrease in a* and b* parameters in fermented cassava ice cream formulated with hydrocolloids (xanthan gum or CMC). The same behaviour was reported by Keshtkaran *et al.* (2013) in milk drinks stabilised with gum tragacanth.

The colour variation (ΔE) in the treated beverages was slightly lower than the control beverage. Nwaokoro and Akanbi (2015) explain that the retention of colour loss in fruit juices is possibly due to the ability of anionic biopolymers to coagulate suspended particles, and control precipitation phenomena and oxidative degradation of pigments. Similarly, Lins *et al.*, (2014) cite that the addition of pectin and gelatin helped to control colour loss in restructured fruits. Martin *et al.*, (2009) state that retention in colour parameter changes is closely related to chitosan treatments in orange juices.

Sensory evaluation: Sensory attributes showed no significant differences as a function of the addition of gums and aloe vera gel ($p>0.05$). Similar results were found in blackberry juices stabilised with xanthan gum and CMC (Akkarachaneeyakorn and Tinrat, 2015). Hamid *et al.* (2014) report that they found no significant differences between carrot-orange mixed nectars formulated with aloe gel and the control in attributes such as appearance, colour and flavour.

Table 3-9 summarises the sensory evaluation for the two types of beverages formulated. No significant difference was estimated in the colour parameter between the beverages, a result in accordance with the instrumental colour evaluation, performed through the CIELAB method. Halim *et al.* (2014) concluded that the addition of xanthan gum, guar and CMC does not affect the colour attribute in cassava ice cream. Also, no significant decrease is reported in relation to the loss of volatile compounds due to the addition of gums or physiologically active components. A similar result is reported in a milk drink stabilised with gum tragacanth (Keshtkaran *et al.*, 2013). On the other hand, a significant decrease in the taste of the beverages associated with the incorporation of gums and aloe vera is observed. This decrease is probably due to the incorporation of aloe gel, which has a low TSS concentration, and may decrease the sweetness in the beverages. *Elbandy et al.* (2014) found similar behaviour in mango nectars with aloe. In turn, the incorporation of gums can reduce the sweetness of beverages. Ibrahim *et al.*, (2011) argue that the addition of xanthan gum, pectin and CMC reduced the flavour of apple juices, associated with a loss of sweetness.

The appearance of the beverages was evaluated in terms of qualities such as homogeneity of the beverage (low sedimentation) and good texture. The increase in the evaluation of the appearance of the treated beverages can be associated with the incorporation of gums. It should be noted that gums have the capacity to increase the viscosity of aqueous suspensions, improve texture and control instability mechanisms such as particle sedimentation; factors that contribute to a better appearance evaluation compared to the control beverage. Similar results have been reported for mixed tomato-carrot, apple and raspberry juices (Ibrahim *et al.,* 2011; Nwaokoro and Akanbi, 2015; Abedi *et al.,* 2014).

CONCLUSIONS

The study has demonstrated the significant effect of hydrocolloids (xanthan gum, CMC) on the degree of stability of tree tomato drinks. The incorporation of xanthan gum was significant ($p<0.05$) in decreasing the settling velocity and increasing the electrostatic repulsion between particles, based on Z-potential values (>30mV). Aloe vera gel intensified the pH and acidity, but did not significantly ($p>0.05$) influence the degree of stability of the beverages.

The mean particle diameters and colour parameters did not vary in relation to the addition of xanthan gum, CMC and aloe vera. The results suggest that due to the hydrophilic and anionic character of the hydrocolloids, the concentrations of xanthan gum and CMC at 0.05% can be considered as suitable treatments for the physical stability of the beverages, without affecting their physicochemical and sensory properties.

ACKNOWLEDGEMENTS

The authors would like to thank the Directorate of Research (DIME) of the National University of Colombia - Medellin, for the financial support of the research project registered under number 19822, through the call "National Programme of Projects for the strengthening of research, creation and innovation in postgraduate studies 2013-2015".

BIBLIOGRAPHY

Abbasi, S., Mohammadi, S. (2013). Stabilization of milk-orange juice mixture using persian gum: Efficiency and mechanism. *Food Bioscience,* 2, 53-60.

Abedi, F., Sani, A. M., Karazhiyan, H. (2012). Effect of some hydrocolloids blend on viscosity and sensory properties of raspberry juice-milk. *Journal of Food Science and Technology,* 51 (9), 22462250.

Akkarachaneeyakorn, S., Tinrat, S. (2015). Effects of types and amounts Ofstabilizers on physical and sensory characteristics of cloudy ready-to-drink mulberry fruit juice. *Food Science & Nutrition,* 3(3), 213-220.

Association of Official Agricultural Chemists (AOAC). Official Methods of Analysis. Association of Official Analytical Chemists. 18th Edition. Arlington, Virginia: AOAC, 2005.

Benitez, E. I., Lozano, J. E. (2007). Effect of gelatin on apple juice turbidity. *Latin American applied research,* 37(4), 261-266.

Bengtsson, H., Hall, C., Tornberg, E. (2011). Effect of physicochemical properties on the sensory perception of the texture of homogenized fruit and vegetable fiber suspensions. *Journal of Texture Studies,* 42, 291-299.

Boghani AH, Raheem A, Hashmi SI (2012) Development and storage studies of blended papayaaloe vera ready to serve (RTS) beverage. *Journal of Food Processing & Technology,* 3(19), 185189.

Bozzi, A., Perrin, C., Austin, S., Arce, V. F. (2007). Quality and authenticity of commercial aloe vera gel powders. *FoodChemistry,* 103(1), 22-30.

Castro, H. I., Benelli, P., Ferreira, S. R., Parada, A. F. (2013). Supercritical fluid extracts from tamarillo *(Solanum betaceum Sendtn)* epicarp and its application as protectors against lipid oxidation of cooked beef meat. *Journal OfSupercritical Fluids,* 76 (2013) 17- 23.

Chaikham, P., Apichartsrangkoon, A. (2012). Comparison of dynamic viscoelastic and physicochemical properties of pressurised and pasteurised longan juices with xanthan addition. *FoodChemistry,* 134(4),2194-2200.

Chatterjee, S., Chatterjee, S., Chatterjee, B. P., Guha, A. K. (2004). Clarification Offruitjuice with chitosan. *Process Biochemistry,* 39, 2229-2232.

Cho, H., Yoo, B. (2015). Rheological characteristics of cold thickened beverages containing xanthan gume based food thickeners used for dysphagia diets. *Journal of the Academy of Nutrition and Dietetics,* 155(1), 106-111.

Delmonte, M. L., Rincon, F., Pinto, G. L., Guerrero, R. (2006). Behavior of the gum from *Enterolobium cyclocarpum* in the preparation peach nectar. *Revista Técnica de la Facultad de Ingeniería - Universidad del Zulia,* 29(1), 23 - 28.

Dluzewska, E., Smiechowski, K., Kowalska, M. (2005). Influence of pH on properties the selected hydrocolloids stabilized model beverage emulsions. *JournalofFood Technology,* 3(4), 542-545.

Elbandy, M. A., Abed, S. M., Gad, S. S., Abdel, F. M. (2014). *Aloe vera* gel as a functional ingredient and natural preservative in mango nectar. *World Journal OfDairy & Food Sciences,* 9 (2): 191-203.

Fasolin, L., and Cunha, R. L. (2012). Soursop juice Stabilizedwith soy fractions: A rheologıal approach. *Ciência e Tecnologia de Alimentos,* 32(3), 558-567.

Fernandez, R., Stinco, C., Hernanz, D., Heredia, F., Vicario, I. (2013). Colourtraining and colour differences thresholds in orange juice. *Food Quality and Preference,* 30(2), 320-327.

Genovese, D. B., Elustondo, M. P., Lozano, J. E. (1997). Colorand cloud stabilization in cloudy apple juice by steam heating during crushing. *JournalofFood Science,* 62, 1171-1175.

Genovese, D. B., Lozano, J. E. (2001). The effect of hydrocolloids on the stability and viscosity of cloudy apple juices. *FoodHydrocolloids,* 15(1), 1-7.

Ghafoor, K., Jung, J. R., Choi, Y. H. (2008). Effects of gellan, xanthan, and λ-carrageenan on ellagic acid sedimentation, viscosity, and turbidity of 'Campbell early' grape juice. *Food Science and Biotechnology,* 17(1), 80-84.

Habeeb, F., Shakir, E., Bradbury, F., Cameron, P., Taravati, M. R., Drummond, A. J., Gray, A. I., Ferro, V. A. (2007). Screening methods used to determine the anti-microbial properties of Aloe vera innergel. *Methods,* 42, 315-320.

Halim, N. R. A., Shukri, W. H. Z., Lanı, M. N., Sarbon, N. M. (2014). Effect Ofdifferent hydrocolloıds on the physicochemical properties, microbiological quality and sensory acceptance of fermented cassava *(Tapai ubi)* ice cream. *InternationalFood Research Journal,* 21(5): 1825-1836.

Hamid, G. H., El-Kholany, E. A., Nahla, E. A. (2014). Evaluation ofAloe vera gel as antioxidant and antimicrobial ingredients in orange-carrot blend nectars. *Middle East Journal of Agriculture Research, 3(4),* 1122-1134.

Homayoonfal, M., Mousavi, F. K. (2015). Modelling and optimising of physicochemical features of walnut-oil beverage emulsions by implementation of response surface methodology: Effect of preparation conditions on emulsion stability. *Food Chemistry,* 174, 649-659.

Huang, X., Kakuda, Y., Cui, W. (2001). Hydrocolloids in emulsions: Particle size distribution and interfacial activity. *FoodHydrocolloids,* 15(4-6), 533-542.

Ibrahim, G., Hassan, I., Abd-Elrashid, A., El-Massry, K., Eh-Ghorab, A., Ramadan, M., Osman, F. (2011). Effect of clouding agents on the quality of apple juice during storage. *Food Hydrocolloids,* 25(1), 91-97.

Colombian Institute of Technical Standards. NTC 3549. Fruit soft drinks. Bogotá: ICONTEC, 1999.

Colombian Institute of Technical Standards. NTC 4105. Fresh fruits. Tree tomato, Specifications. Fruit drinks. Bogotá: ICONTEC, 1997.

Kaneiwa, M., Augusto, P., Cristianini, M. (2013). Effect of high-pressure homogenization (HPH) on the physical stability Oftomatojuice. *Food Research International,* 51(1), 170-179.

Keshtkaran, M., Mohammadifar, M. A., Asadı, G. A., Nejad, R. A., Balaghi, S. (2013). Effect of gum tragacanth on rheological and physical properties of a flavored milk drink made with date syrup. *JournalofDairyScience,* 96(8), 4794-4803.

Koyama, M., Kitamura, U. (2014). Development of a new rice beverage by improving the physical stability of rice slurry. *Journal ofFood Engineering,* 131,89-95.

Leiberman, H. A., Reiger, M. M. and Banker, G. S. Pharmaceutical Dosage Forms: Disperse Systems. Second Edition. NewYork, USA: Marcel Dekker, 1998, 559p.

Leite, T. S., Augusto, P. E., Cristianini, M. (2014). The use of high-pressure homogenization (HPH) to reduce consistency of concentrated orange juice (COJ). *Innovative Food Science and Emerging Technologies,* 26, 124-133.

Liang, C., Hu., X., Ni, Y., Wu., J., Chen, F., Liao, X. (2006). Effect of hydrocolloids on pulp sediment, white sediment, turbidity and viscosity of reconstituted carrot juice. *Food Hydrocolloids,* 20(8), 11901197.

Lins, A. C., Cavalcanti, B. D., Azoubel, P. M., Melo, E. A., Maciel, S. M. (2014). Effectofhydrocolloids on the physicochemical characteristics of yellow mombin structured fruit. *Food Science and*

Technology, 34(3): 456-463.
McClements, D. J. Food emulsions: Principles, practice, and techniques. Second Edition. Boca Raton, USA: CRC Press, 2005, 632 p.
Marquez, C., Otero, C., Cortes, M. (2007). Physiological, textural, physiological, physicochemical and microstructural changes of tree tomato *(Cyphomandra betacea S.)* in postharvest. *Journal of the Faculty of Pharmaceutical Chemistry,* 14 (2), 9-16.
Marquez, Carlos (2009). Physiological, physicochemical, rheological, nutraceutical, structural and sensory characterisation of soursop *(Annona muricata L. cv. Elita).* Doctoral thesis. National University of Colombia, Faculty of Agricultural Sciences, p. 274 p.
Martin, D. A., Rico, D., Barat, J. M., Barry, R. C. (2009). Orange juices enriched with chitosan: Optimisation for extending the shelf-life. *Innovative Food Science and Emerging Technologies*, 10, 590-600.
Meza, N., Manzano, M. J. (2009). Tree tomato *(Cyphomandra betaceae [Cav.] Sendtn)* fruit characteristics based on aril coloration in the Venezuelan Andean Zone. *Revista UDO Agricola,* 9(2), 289-294.
Mirhosseini, H., Tan, C. P. (2010). Effect of various hydrocolloids on physicochemical characteristics oforange beverage emulsion. *JournalofFood, Agriculture & Environment,* 8(2), 308 - 313.
Mirhosseini, H., Tan, C. P., Aghlara, A., Hamid, N. S., Yusof, S., Chern, B. H. (2008). Influence of and CMC on physical stability, turbidity loss rate, cloudiness and flavor release of orange beverage emulsion during storage. *Carbohydrate Polymers,* 73, 83-91.
Moraes, I. C. F. F., Fasolin, L. H., Cunha, R. L., Menegalli, F. C. (2011). Dynamic and steady-shear rheological properties of xanthan and guar gums dispersed in yellow passion fruit pulp *(Passiflora edulis f. flavicarpa). Brazilian Journal of Chemical Engineering,* 28(3), 483 - 494.
Nascimento, G. E., Hamm, L. A., Baggio, C. H., Paula, W. M., Iacomini, M., Cordeiro, L. M. (2013). Structure of a galactoarabinoglucuronoxylan from tamarillo *(Solanum betaceum),* a tropical exotic fruit, and its biological activity. Food Chemistry, 141,510-516.
Nwaokoro. O, G., Akanbi, C. T. (2015). Effect of the addition of hydrocolloıds to tomato-carrot juice blend. *Journal OfNutritional Health & Food Science,* 3(1): 1-10.
Paquet, E., Alexandra Bedard, A., Lemieux, S., Turgeon, S. L. (2014). Effects of apple juice-based beverages enriched with dietary fibres and xanthan gum on the glycemic response and appetite sensations in healthy men. *Bioactive Carbohydrates and DietaryFibre,* 4, 39-47.
Qin L, Xu S, Zhang W. (2005). Effect of enzymatic hydrolysis on the yield of cloudy carrot juice and the effects of hydrocolloids on colour and cloud stability during ambient storage. *Journal of Science OfFoodandAgriculture,* 85, 505-512.
Ramachandra, C.T., and Srinivasa, P. (2008). Processing of Aloe vera gel: A review. *American Journal OfAgricultural and Biological Sciences,* 3(2), 502-510.
Rungsardthong, V., Wongvuttanakul, N., Kongpien, N., Chotiwaranon, P. (2006). Application of fungal chitosan for clarification of apple juice. *Process Biochemistry,* 41,589-593.
Sahin, H., Ozdemir, F. (2007). Effect of some hydrocolloids on the serum separation of different formulated ketchups. *JournalofFoodEngineering,* 81,437-446.
Schramm, L. L. Emulsions, Foams, and Suspensions: Fundamentals and Applications. In: Schramm, L. L. Colloid Stability. Weinheim: Wiley-VCH, 2005, 117-152 p.
Sherafati, M., Kalbası-ashtarı, A., Ali-Mousavi, S. M. (2013). Effects of low and high acyl gellan gums on engineering properties of carrot juice. *JournalofFood Process Engineering,* 36(4), 418-427.
Shubhra, B., Swatı, K., Pushpinder, S. R., Savita, S. (2014). Studies on Aloejuice supplemented kinnow nectar. *Research Journal OfAgriculture and Forestry Sciences,* 2(8), 14-20.
Silva, V. M., Kawazoe, S. A., Barbosa, G., Dacanal G., Ciro-Velasquez, H. C., Cunha, R. L. (2010). The effect Ofhomogenisation on the stability Ofpineapple pulp. *International Journal of Food Science and Technology,* 45(10), 2127-2133.
Yaron, A., Cohen, E., Arad, S. (1992). Stabilization of Aloe vera gel by interaction with sulfated polysaccharides from red microalgae and with xanthan gum. *Journal of Agricultural and Food Chemistry,* 40(8), 1316-1320.

3 Chapter

EFFECT OF HYDROCOLOIDESYALOE VERA *(Aloe barbadensis)* INCORPORATION ON THE RHEOLOGICAL PROPERTIES OF TOMATO TREE DRINKS *(Cyphomandra betacea)*

ABSTRACT - The effect of the incorporation of hydrocolloids (xanthan gum, CMC) and aloe vera *(Aloe barbadensis Miller)* on the rheological characteristics of tree tomato *(Cyphomandra betacea)* beverages was evaluated by rotational and oscillatory tests. The results obtained were fitted to various rheological models, and the Arrhenius equation was used to study the dependence of rheological parameters on temperature. The results showed that for the range of concentrations and temperatures studied, the formulated beverages show a pseudoplastic behaviour where the Power Law model showed the best fit to the experimental data. The oscillatory tests reveal the predominance of the elastic modulus (G'>G") over the whole frequency range. The estimated parameters of the oscillatory rheological study did not show a linear relationship, based on the Arrhenius principle. Concentrations of xanthan gum ≥ 0.025% and CMC ≥ 0.05% are assessed as the best treatments, associated with the increase of the consistency coefficient *(K)* and the trend of the loss tangent (δ) values which are assumed to be excellent indicators of stability in suspensions.

Keywords: Rheology, stability, hydrocolloids, viscosity, pseudoplastic.

ABSTRACT - The effect of the incorporation of hydrocolloids (xanthan gum, CMC) and aloe vera *(Aloe barbadensis Miller)* on the rheological characteristics of tree tomato drinks *(Cyphomandra betacea)* was evaluated by rotational and oscillatory tests. The results were adjusted to different rheological models, and the Arrhenius equation was used to study the dependence of the rheological parameters with temperature. The results showed that for the range of concentrations and temperatures studied drinks made show shear thinning where the power law model showed the best fit with respect to the experimental data. The oscillatory tests reveal the predominance of elastic modulus (G' > G^{1} ') throughout the frequency range. The estimated parameters of oscillatory rheological study showed no linear relationship based on the principle of Arrhenius. Xanthan gum concentrations ≥ ≥ 0.025% and 0.05% CMC are measured as the best treatments, associated with increased consistency coefficient (K) and the trend of the values of the loss tangent (δ) which are assumed as excellent suspension stability indicators.

Keywords: Rheology, stability, hydrocolloids, viscosity, pseudoplastic.

INTRODUCTION

The tree tomato *(Cyphomandra betacea*) is a plant belonging to the Solanaceae family, and its cultivation is of great socioeconomic importance in the Colombian Andean region. It is a native plant of South America, and is appreciated for its succulent pulp, pleasant aroma, sweet and sour taste and orange colour (Meza and Manzano, 2009). The tree tomato is considered an exotic fruit, with excellent nutritional characteristics as a source of β-carotene, pyridoxine, ascorbic acid, magnesium, potassium, phosphorus, calcium, free amino acids, high fibre content and low caloric intake (Márquez *et al.*, 2007; Kou *et al.*, 2009; Vasco *et al.*, 2009). The pulp is used in the production of juices, nectars, jams, salads, preserves and dehydrated flakes (Meza and Manzano, 2009).

The incorporation of tropical fruits in juice production is an alternative to reduce post-harvest losses, reduce production surpluses and add value to raw materials. A fruit juice is a multiphase system that presents a large amount of insoluble material, which tends to precipitate leading to phase separation after a certain storage time (Genovese, 1997). The addition of aloe vera (*Aloe barbadensis* Miller) can affect the stability of the juices, as it is a mucilaginous gel consisting of carbohydrates, amino acids, lipids, sterols, minerals and vitamins (Choi and Chung, 2003; Hamman, 2008; Domínguez *et al.*, 2012). Studies highlight the functional properties of aloe vera, and mention that the inclusion of these components in food strengthens its nutritional value (Habeeb *et al.*, 2007; Ramachandra *et al.*, 2008; Boghani *et al.*, 2012).

Hydrocolloids are polysaccharides or substances of high molecular weight, used in food suspensions as stabilisers, because they increase the viscosity of the continuous phase and lead to ionisation of particles in aqueous solutions (Genovese and Lozano, 2006). Although they are added in small concentrations (less than 1%), they have a significant influence on the rheological and sensory properties of foods. Several investigations have studied the effect of xanthan gum, carboxymethyl cellulose (CMC), guar gum, gellan gum, chitosan, pectin and modified starches on the stability, rheological behaviour and sensory properties of fruit juice (Genovese and Lozano, 2001; Liang *etal.,* 2006; Meng and Rao 2005; Ibrahim *et al.,* 2011; Paquet *etal.,* 2014).

Theological characterisation is important for equipment design, heat transfer, transport system, process optimisation and quality assessment (Dak *et al.,* 2006). Different mathematical models have been used to describe food flow behaviour. The simplest type of rheological behaviour is Newtonian, in which there is a linear relationship between shear stress and strain rate. However, most fluid feeds do not exhibit this simple behaviour, which requires more complex models for characterisation. The most commonly used models are the Bingham, Power Law, Herschel-Bulkley and Casson models. The power law model is the most widely used to describe the rheological behaviour of most fruit juices, especially in handling operations, as it is convenient, simple and easy to use (Gratáo *et al.,* 2007; Ahmed *et al.,* 2007; Quek *et al.,* 2013). Also, rheological behaviour is influenced by temperature (Falcone *et al.,* 2007), where the Arrhenius relation has often been used to describe the effect of temperature on various theological parameters.

However, many theological phenomena cannot be described in terms of viscosity alone, and therefore elastic behaviour must be evaluated. Viscoelastic testing of macromolecular dispersions can be determined by dynamic testing using oscillatory sweeps. In a frequency oscillatory test, the sinusoidal deformation cycle is used to determine the storage or elastic modulus (G'), loss or viscous modulus (G") and loss tangent (tan δ). Oscillatory tests provide theological parameters, without altering the internal network structures of the materials used (Ahmed *et al.,* 2007). Viscoelastic properties are very useful in the design and prediction of product stability. Thus, the study and description of viscoelastic properties of liquid foods is important for a better understanding of their behaviour during processing, storage and consumption (Augusto *et al.,* 2013a).

The aim of this research was to study the effect of the addition of hydrocolloids and aloe vera on the theological behaviour of tree tomato drinks, and to estimate their effect on the degree of suspension stability.

.

MATERIALS AND METHODS

Tree tomato *(Cyphomandra betacea S.)* fruits of the common orange variety were used. The fruits were selected with a maturity grade, colour typology 6 (NTC 4105/1997). Hydrocolloids (xanthan gum, sodium carboxymethyl cellulose-CMC) and food grade preservatives (sodium benzoate and potassium sorbate) supplied by Bell Chem International S.A. The food grade Aloe vera gel with 98% purity was purchased from Ayala Benard S.A.S.

Preparation of the nectar drink. The plant material was disinfected in 100 ppm sodium hypochlorite solutions for 10 minutes, and finally washed in water. The pulp was extracted in a pulping machine (D1000, CITALSA, Colombia). The nectars were formulated from a 300g base, with a pulp content of 18% (NTC 3549/1999). Sucrose was used as sweetener, and water was added as solvent until a beverage with a soluble solids concentration of 10° Brix was obtained. The respective amount of hydrocolloids was added after preparation in a suspension of water at 40° C. In addition, a preservative mixture equal to 0.125% was added to prolong the shelf life and to study the degree of stability under storage conditions. The beverages were homogenised in a ULTRA-TURRAX disperser (IKA, T25 Basic, Germany) for 60 seconds at 5000 rpm. They were also subjected to a mild pasteurisation process at

60°C for one minute. The final product was packed in plastic containers (PET) and stored under refrigeration at 4.0° C until further characterisation.

Rotational tests. The flow curves were determined in a viscometer (Brookfield, Model DV III Ultra, USA) in rotational mode, using concentric cylinder geometry (SC4-21, 2.5 cm diameter). In order to eliminate the possible thixotropy presented by the product, strain rate (y) sweeps were performed in the range from 0 to 200 s^1 , first in an ascending, then in a descending and finally in an ascending way. The curve obtained from the last procedure was taken as a reference to study the theological behaviour of the formulated beverage. The equipment directly recorded the shear stress (σ), strain rate (y) and apparent viscosity (μ). The experimental values of the flow curves were fitted to Newtonian, Power Law, Herschel-Bulkley and Casson rheological models (Table 4-1). The experimental data fit was evaluated using the non-linear regression statistical tool, and parameters such as regression coefficient $(R\ )^2_1$ mean squares of the error (MSE) and Fisher's test for model significance were determined.

Table 3-1: Rheological models studied.

Model	Equation	Parameters
Newton	$\sigma = K(\gamma)$	K
Power Lawyer	$\sigma = K(\gamma\ )^n$	K,n
Herschel-Bulkley	$\sigma\text{-}\sigma_0 + K(\gamma\ )^n$	$\sigma_0\ ,K,n$
Casson	σ0.5=)0.S+K (y 0.S)	$\sigma_0\ ,K$

Experimental measurements were carried out at temperatures of 10, 20, 30, 40 and 50°C. The Arrhenius equation (Eq. 4.1) was used to evaluate the effect of temperature on the rheological parameters.

$$X = X_o\, e^{\left(E_a/_{RT}\right)} \qquad \textbf{(Ec. 4.1)}$$

To obtain the constant, X_0 and the activation energy, E_a , the Arrhenius equation was linearised as a function of temperature (T). The linear regression tool was used, determining the regression coefficient and the significance of the model through Fisher's test.

Oscillatory tests. The viscoelastic behaviour of the material was determined on a rheometer (Anton Paar, MCR 302, Austria) in oscillatory mode, using the geometry of parallel plates with a diameter of 25 mm (P-PTD 200/E, Anton Paar, MCR 302). A gap of 1.0 mm between plates was set. The rheometer was complemented with a Peltier system for sample temperature control, which was previously controlled and monitored. The perimeter of the exposed sample was covered with a metal chamber to minimise evaporation at high temperatures. Initially, an amplitude sweep was performed at 0-100% strain and 1.0 Hz frequency to determine the linear viscoelasticity region. The viscoelastic characterisation of the material was obtained by varying the frequency from 0.01 to 100 Hz (Augusto *et al.*, 2013b), where all rheological measurements were carried out in duplicate. The storage modulus (G') a measure of the elastic property, loss modulus (G") a measure of the viscous property, and loss tangent (tan δ) were obtained directly from the Rheocompass software version 1.12 (Anton Paar, Austria). The behaviour of the above-mentioned variables was studied as a function of temperature at 10, 20, 30, 40 and 50°C.

Experimental design. A central rotational design with axial points was established for the experimentation, with four replicates at the central point. The factors and levels established in the study are defined in Table 4-2. The results were statistically analysed by generating response surfaces, analysis of variance, lack of fit test and determination of regression coefficients using Statgraphics Centurion XVI software, version 16.1.18.

Table 3-2: Factors analysed and coding of the levels established.

Factor	Symbol	Under	Medium	High

Coding		-1	0	+1
Aloe vera (%)	X_1	0,5	1,0	1,5
Xanthan gum (%)	X_2	0,025	0,05	0,075
Sodium carboxymethyl cellulose - CMC (%)	X_3	0,025	0,05	0,075

The experimental data were fitted to the following second-order model (Eq. 4.2):

$$Y - \beta o + \sum L\ ßiX\text{-}í + \sum L\ ßn\chi^2\ ı + \sum\sum F<\surd=ı\ ßijXiXj$$

Where β_0 , $ßi$, β_u , $ßij$ are the regression coefficients for the intercept, linear and quadratic interaction terms, respectively, and X is the independent variable.

ANALYSIS OF RESULTS

Rotational rheology: The experimental data of the rotational measurements of tree tomato beverages were fitted to several rheological models such as Newton, Power Law, Herschel-Bulkley and Casson. The models were found to be statistically significant ($p<0.05$) with a regression coefficient greater than 0.9 (Table 4-3). In the Herschel-Bulkley model relating three parameters, negative values for the creep threshold were estimated which do not make sense from a physical point of view. However, the power law is the model that best fits the experimental data with $R^2 \approx 1$, and low error mean squares (EQS) values. Several studies have explained the flow behaviour in fruit juices using the power law model (Cabrai *et al.*, 2007; Chin *et al.*, 2009; Sogi *et al.*, 2010; Goula and Adamopoulos, 2011; Quek *et al.*, 2013).

Table 3-3: Parameters of rheological models estimated on tree tomato beverages under temperature conditions at 30 C.°

Model	Parameters	Control	Point Minimum	Point Medium	Point Maximum
Newton	*K*	0,006	0,010	0,013	0,018
	R^2	0,924	0,916	0,900	0,849
	GCE	0,012	0,045	0,076	0,228
Power Law	*K*	0,032	0,065	0,089	0,177
	n	0,677	0,647	0,624	0,564
	R^2	0,997	0,999	0,999	0,999
	GCE	0,004	0,001	0,008	0,011
Herschel-Bulkley	*K*	0,028	0,063	0,086	0,179
	σo	0,023	0,015	0,019	-0,005
	n	0,699	0,654	0,631	0,563
	R^2	0,998	0,999	0,999	0,999
	GCE	0,004	0,001	0,008	0,013
Casson	*K*	0,065	0,087	0,096	0,098
	σo	0,027	0,048	0,058	0,084
	R^2	0,979	0,975	0,964	0,954
	GCE	0,002	0,005	0,009	0,012

The flow curves show a decrease in shear stress as the strain rate increases, expressed as a decrease in the apparent viscosity of the different formulated tree tomato drinks (Fig. 4-1). This behaviour is due to non-Newtonian shear thinning, characteristic of a pseudoplastic fluid. Abbasi and Mohammadi (2013) reported pseudoplastic behaviour, in orange juices by evaluating different concentrations of hydrocolloids. Banerjee and Ghosh (2015) also report that tamarind juice stabilised with xanthan gum and CMC at various temperatures behaved as a pseudoplastic fluid.

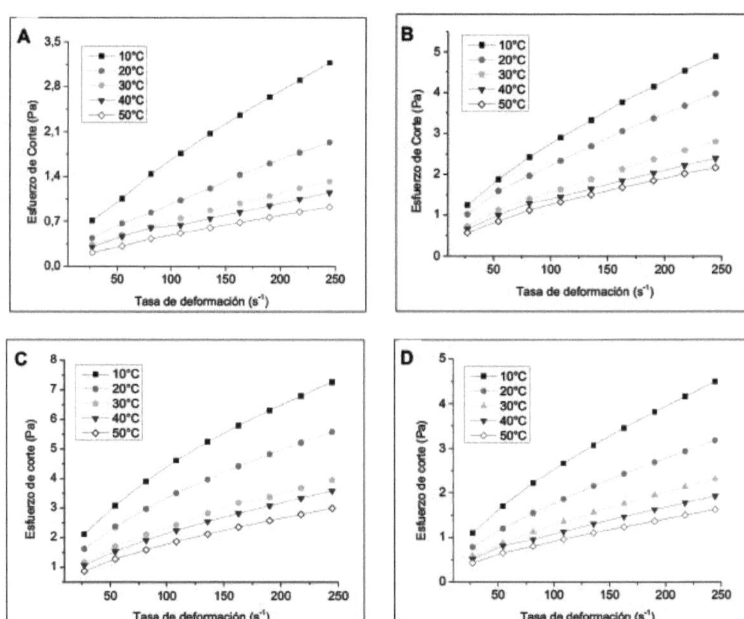

Figure 3-1: Rheological behaviour of tree tomato drink at various temperatures: (A) Control, (B) Minimum point, (C) Mid-point and (C) Maximum point.

Table 4-4 shows the values of the consistency coefficient, *K*, adjusted to the power model. A significant increase ($p<0.05$) is observed in the values of the consistency coefficient of the hydrocolloid-formulated beverages with respect to the control. In other words, there was an increase in the apparent viscosity of the beverages. Some authors have cited that the addition of xanthan gum and CMC increased the viscosity of the continuous phase in carrot and apple juices with respect to the control, improving particle stability in the dispersion (Liang *et al.,* 2006; Ibrahim *et al.,* 2011). The coefficient *K*, in the power law model ranged from 0.019 to 0.326 $Pa.s^{n}$. Similar results have been reported for cherry, grapefruit and guava juices (Cabrai *et al.,* 2007; Chin *et al.,* 2009; Fasolin et *al.*
Cunha, 2012). Additionally, there is a marked effect on the consistency index with respect to temperature. This behaviour indicates a tendency towards a decrease in viscosity and an increase in the rate of fluidity of the beverages at higher temperatures. This is probably due to the fact that cohesion and the rate of transfer of the amount of molecular motion in suspensions decreases with increasing temperature, decreasing the shear strength of the fluid (Earle, 1985). Xu *et al.* (2013) report that xanthan gum solutions decrease in apparent viscosity with increasing temperature. Similar behaviour has been described in carrot, watermelon, tamarind and grapefruit juices (Vandresen *et al.,* 2009; Sogi *et al.,* 2010; Ahmed *et al.,* 2007; Chin *et al.,* 2009).

Table 3-4: Consistency coefficient *K* ($Pa.s^{n}$) as a function of temperature and fit to the Arrhenius model.

Essay	X_1	X_2	X_s	Temperature (° C)					Arrhenius parameters	
				10	20	30	40	50	K_0 ($Pa.s)^{n}$	E_a (kJ/mol)
1	0	0	1,68	0,138	0,097	0,084	0,092	0,071	1,29E-03	10,824
2	0	-1,68	0	0,073	0,051	0,044	0,053	0,035	2,83E-04	12,871
3	-1,68	0	0	0,111	0,092	0,072	0,086	0,065	2,46E-03	8,914
4	-1	1	1	0,326	0,233	0,179	0,146	0,115	8,37E-05	10,384

5	1	-1	1	0,084	0,063	0,044	0,046	0,039	1,78E-04	10,318
6	0	0	0	0,149	0,127	0,090	0,095	0,075	6,15E-04	14,556
7	1	1	-1	0,163	0,148	0,183	0,109	0,078	6,15E-02	16,267
8	0	0	0	0,146	0,124	0,088	0,093	0,074	6,93E-03	12,544
9	1	-1	-1	0,093	0,069	0,073	0,048	0,040	1,31E-04	16,465
10	0	0	-1,68	0,100	0,082	0,069	0,067	0,046	3,46E-04	15,365
11	0	1,68	0	0,271	0,232	0,183	0,185	0,162	9,49E-02	11,582
12	1,68	0	0	0,178	0,136	0,099	0,108	0,087	7,54E-04	12,690
13	0	0	0	0,150	0,129	0,092	0,098	0,077	7,89E-04	14,321
14	0	0	0	0,106	0,115	0,090	0,095	0,074	2,94E-02	14,600
15	-1	-1	1	0,175	0,122	0,084	0,068	0,052	1,64E-05	9,933
16	1	1	1	0,326	0,244	0,177	0,164	0,131	2,36E-02	16,923
17	-1	-1	-1	0,131	0,090	0,065	0,065	0,053	1,18E-04	16,278
18	-1	1	-1	0,226	0,177	0,147	0,119	0,106	4,51E-04	14,585
Control	-	-	-	0,063	0,035	0,032	0,037	0,019	1,02E-06	20,330

The regression coefficients were greater than 0.90.

The values of the flow behaviour index are below unity (*n<1*), indicating the pseudoplastic behaviour of the tree tomato drinks (Table 45). A decrease in the values of the index, *n*, is observed in the nectars formulated with hydrocolloids, with respect to the control. The reduction of pseudoplasticity in the beverages is possible due to the incorporation of hydrocolloids in the suspension. Xu *et al.* (2013) mention that hydrocolloids in general present a non-Newtonian behaviour of pseudoplastic type, given by a decrease in apparent viscosity at high shear rates. The index *n,* for the power law model ranged from 0.522 to 0.728 Pa.s$^{\pi}$. Cabrai *et al.*, (2007) and Quek *et al.*, (2013) report similar values in cherry and soursop juices, respectively, formulated at a solids concentration at 20° Brix. The flow behaviour index (*n*) tends to decrease with increasing temperature, based on the significant fit to the Arrhenius relation. The decrease in the *n* index intensifies the pseudoplasticity of the tree tomato beverages. This coincides with the
reported by Kumar and Kumar (2015), who highlighted an increase in Pseudoplasticity in beetroot juices with a solids concentration of 25°Brix, in the range of 30 to 80° C. Yogurtçu and Kamişli (2006), also report a reduction in the rate of flow behaviour as temperature increases, in concentrated fruit syrups.

Table 3-5: Flow behaviour index (*n*), as a function of temperature and fit to the Arrhenius model.

Temperature (° C)				Arrhenius parameters						
Essay	Xi	X_2	X_3	10	20	30	40	50	*Пo*	E_a (kJ/mol)
i	0	0	1,68	0,639	0,659	0,632	0,584	0,615	0,242	2,442
2	0	-1,68	0	0,703	0,724	0,690	0,625	0,675	0,160	0,627
3	-1,68	0	0	0,645	0,638	0,629	0,576	0,610	0,422	0,006
4	-1	1	1	0,566	0,561	0,561	0,568	0,573	0,416	0,213
5	1	-1	1	0,710	0,687	0,697	0,642	0,637	0,282	2,180
6	0	0	0	0,634	0,625	0,624	0,584	0,612	0,486	0,651
7	1	1	-1	0,615	0,575	0,501	0,550	0,577	0,444	0,675
8	0	0	0	0,634	0,626	0,623	0,586	0,609	0,38	0,652
9	1	-1	-1	0,687	0,658	0,595	0,626	0,628	0,307	2,874
10	0	0	-1,68	0,664	0,625	0,614	0,601	0,631	0,300	1,829
11	0	1,68	0	0,549	0,548	0,546	0,522	0,535	0,441	0,524
12	1,68	0	0	0,618	0,623	0,622	0,578	0,602	0,428	0,936
13	0	0	0	0,637	0,625	0,622	0,583	0,611	0,474	0,688
14	0	0	0	0,605	0,626	0,621	0,583	0,613	0,489	0,609
15	-1	-1	1	0,653	0,660	0,664	0,664	0,677	0,287	0,619
16	1	1	1	0,565	0,569	0,564	0,560	0,567	0,488	0,366
17	-1	-1	-1	0,643	0,648	0,647	0,613	0,620	0,401	1,160

18	-1	1	-1	0,581	0,572	0,555	0,563	0,550	0,387	0,952
Control	-	-	-	0,713	0,728	0,677	0,722	0,704	0,143	3,922

The regression coefficients were greater than 0.90.

The Arrhenius relation was used to describe the effect of temperature on the rheological parameters estimated in the power law model, obtaining significant models for the consistency coefficient (K_0). Table 4-4 details the results of the frequency factor K_0 and activation energy (E_a) determined from the Arrhenius equation. It was estimated that xanthan gum concentration exerted a significant effect ($p<0.05$) on the frequency factor, K_0 . The factor K_0 , ranged from 1.02-06 to 9.49E-02 values close to those reported by Quek *et al.,* (2013) in soursop juice concentrates in a temperature range of 10 to 50°C. Goula and Adamopoulos (2011) estimated similar values of the frequency factor in kiwifruit juices with a solids concentration of 15 and 30°Brix, for a temperature range of 25 to 65°C. The mathematical model describing the behaviour of the *K-factor*$_0$, presented a regression coefficient of 0.79 (Table 4-6), and the test of fit shows that the selected model is adequate to describe its behaviour ($p>0.05$). In Fig. 4-2, an increase in the *K-factor*$_0$ is observed as the xanthan gum concentration increases. Garcia *et al.,* (2000) cite that xanthan gum is a high molecular weight stabiliser, which generates a significant increase in apparent viscosity and decrease in flow rate in particle suspensions, compared to the effect of other gums. In addition, no significant effect was found due to the incorporation of aloe vera and CMC. However, as the CMC concentration increased, the *K-factor*$_0$ decreased, possibly due to the effect of temperature on the stability of the hydrocolloid. Coffey *et al.,* (2006) argue that viscous CMC solutions are affected by temperature, gradually decreasing with increasing thermal effect.

Table 3-6: Regression parameters for consistency coefficient *K*, flow rate n_0 , consistency factor (K_0), flow rate factor (n_0) and activation energy (E_a).

Regression coefficients	tf(Pa.s^n), [10°C].	K_0 (Pa.s)^n	*П*, [10° C]	n_0
ßo	0,26	-0,01	0,64	0,21
ßl	-0,15	0,02	0,06	-0,16
β_2	-3,82	-1,69	0,24	5,39
β_3	-1,81	1,36	-0,05	5,09
βn	0,66	0,84	NS	NS
βl2	0,10	-0,37	-0,22	1,04
β 13	45,6	NS	NS	NS
β_{22}	0,05	-0,02	NS	NS
$\beta\ 3_2$	34,92	19,39	-4,09	-70,68
$\beta\ 3_3$	NS	NS	10,33	-87,37
R^2	0,82	0,79	0,91	0,87
Model (p-value)	0,03	0,05	0,01	0,01
Missing fit (p-value)	0,06	0,31	0,26	0,58

*NS: Coefficients not significant

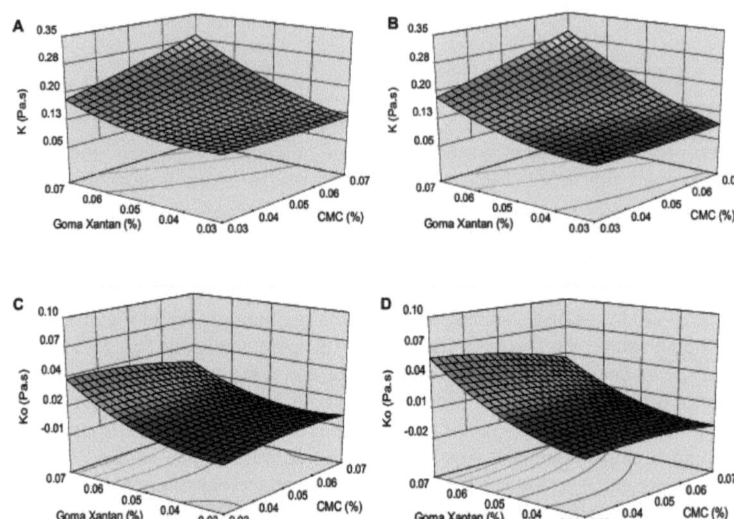

Figure 3-2: Response surfaces for consistency coefficient *(K)* and frequency factor (K_0): A) *Ka* 10° C and 0.5% aloe, B)A) *Ka* 10° C and 1.5% aloe; C) K_0 for 0.5% aloe, D) K_0 for 1.5% aloe.

On the other hand, the xanthan gum concentration significantly influenced ($p<0.05$) the behaviour of the factor n_0 (Table 4-5), showing a gradual increase due to the hydrocolloid concentration (Fig. 4-3). In other words, the tendency of the beverages towards pseudoplastic behaviour is evident. Williams and Phillips (2009) argue that the transition between Newtonian and pseudoplastic behaviour in some beverages is associated with the increase in viscosity provided by high molecular weight hydrocolloid solutions. However, at higher concentrations of xanthan gum, there is a slight decrease in the *n-factor*$_0$. This is probably associated with conformational changes in the hydrocolloid structure with increasing temperature, causing a greater disorganisation of the molecules and a decrease in the viscosity of the suspension (Xu *et al.,* 2013). The estimated mathematical model describing the behaviour of the *n* factor$_0$ recorded a coefficient of determination, R^2 of 0.86 (Table 4-6). Likewise, the model obtained in relation to the goodness-of-fit test showed a good statistical estimate ($p>0.05$). The value of the factor n_0 , for the adjusted flow rate at various temperatures according to the Arrhenius equation ranged from 0.16-0.49 (Table 4-5). These values are higher than those reported by Vandresen *et al.* (2009) for pasteurised carrot juice with a total solids concentration of 8%. This difference may be related to the effect of the pasteurisation process, the range established to evaluate the thermal effect, incorporation of hydrocolloids and solids concentration.

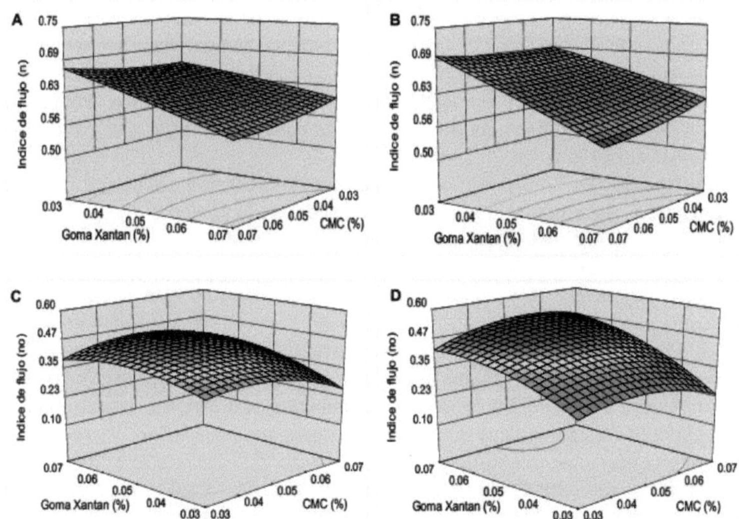

Figure 3-3: A) Response surfaces for flux rate (*n*) and frequency factor (n_0): A) *n* a10° C for 0.5% aloe, B) *n* a10° C for 1.5% aloe, C) n_0 for 0.5% aloe; D) n_0 for 1.5% aloe.

Given the importance of evaluating the stability of the beverages under refrigerated storage conditions, the rheological parameters *Kyn* were analysed at a temperature of 10° C. The xanthan gum concentration was found to have a significant effect ($p<0.05$) on the rheological parameters, with increased beverage consistency and a tendency towards a pseudoplastic fluid as the concentration of the hydrocolloid increased (Fig. 4-2, Fig. 4-3). Whistler and BeMiller, (1997) report that xanthan gum solutions form rigid structures due to the formation of high molecular weight aggregates, giving the suspensions increased viscosity and remarkable pseudoplasticity. Similar results have been reported in passion fruit pulp stabilised with xanthan gum (Moraes *et al.,* 2011). In addition, the increase in apparent viscosity and greater stability of suspensions as xanthan gum concentration is increased is highlighted in apple juice (Genovese and Lozano, 2001) and reconstituted carrot juice (Liang *et al.,* 2006). Stokes' law illustrates that sedimentation of particles in suspensions is inversely proportional to their viscosity, which means that higher viscosity can beneficially increase juice stabilisation (Genovese and Lozano, 2001). The mathematical models estimated for *Kyna* 10° C, presented coefficients of determination of 0.82 and 0.91, respectively (Table 4-6). Likewise, the fit test allows us to infer that the estimated coefficients satisfactorily describe the behaviour of the variables analysed ($p>0.05$).

On the other hand, it was found that the concentrations of gum and aloe vera had no significant effect ($p>0.05$) on the activation energy parameter, E_a . The activation energy (E_a) is the threshold energy that must be exceeded for the elemental flow process to occur (Shamsudin *et al.,* 2013). The value of E_a estimated from the consistency coefficients characteristic of the various formulated beverages ranged from 8.914 to 20.330 kJ/mol (Table 4-4). These results are below those estimated for grapefruit (Quek *et al.,* 2013) and soursop (Chin *et al.,* 2009) juice concentrates with a solids content of 20° Brix. These differences may be marked by the temperature ranges established in each study. Furthermore, Haminuik *et al.,* (2006) state that the most significant changes in viscosity are reached at high temperatures, and are associated with high E values$_a$. However, a decrease in E_a is detected in the hydrocolloid-formulated beverages compared to the control. This decrease is possibly due to the incorporation of hydrocolloids such as xanthan gum and CMC, substances that in suspension acquire a pseudoplastic behaviour, characterised by a decrease in apparent viscosity at high strain rates. According to Krokida *et al.* (2001), the value of the activation energy tends to increase in those suspensions where the flow behaviour is close to Newtonian, as simulated by the control beverage, which showed a flow behaviour index closer to unity.

Oscillatory Rheology - The elastic or storage modulus (G') and viscous or loss modulus (G") as a function of frequency in tree tomato beverages are plotted in Fig. 4-4. The elastic modulus (G') always predominated over the viscous modulus (G"), for all formulations. This indicates that the elastic properties were dominant over the viscous ones, therefore, the beverages can be defined as weak gels. This behaviour is present in most suspensions with a network structure, characteristic of products derived from vegetables and fruits (Augusto *et al.,* 2013b). The same authors report the predominance of the elastic component in tomato juices homogenised at high pressures. Moraes *et al.* report a gel-like behaviour (G'>G") in passion fruit juices stabilised with xanthan. Benchabane and Bekkour (2008) highlight the dominant effect of elastic modulus compared to viscous modulus in low concentrated CMC suspensions (<2.0%). Tiziani and Vodovotz (2005) report similar results in tomato juices to which soy protein was added. The authors consider that the formulated product physically behaves as a weak gel (G'>G"), due to the cross-linked structure formed by the highly esterified pectin in the presence of calcium ions. Moelants *et al.,* (2013) cite that carrot-derived suspensions behave elastically when evaluated in an angular frequency range of 0-10 rad/s.

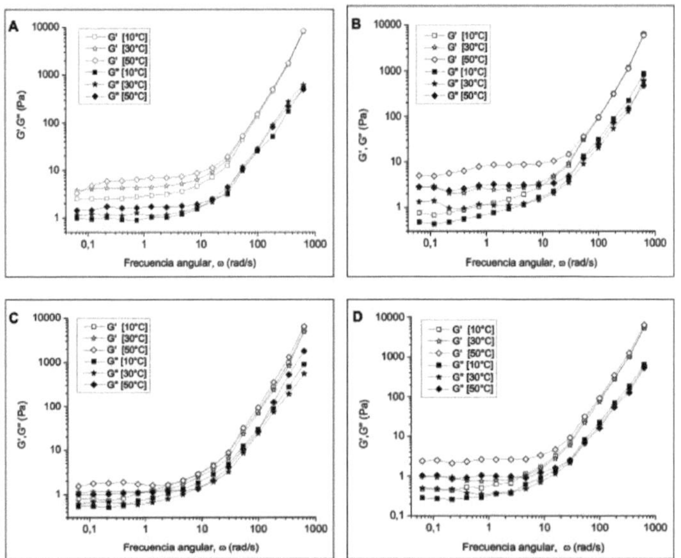

Figure 3-4: Viscoelastic behaviour of tree tomato beverages at temperatures of 10, 30 and 50° C: (A) Control, (B) Minimum point, (C) Centre point, and (D) Maximum point.

The results of the frequency sweeps were analysed using a power law model (Eq. 3 and 4) where *K", K", n', n"* are constants and ω is the angular frequency. The estimated parameter values are summarised in Table 4-7.

$$G' = K'(\omega)^{n'} \qquad \textbf{(Ec. 4.3)}$$

$$G'' = K''(\omega)^{n''} \qquad \textbf{(Ec. 4.4)}$$

The values of *K'* and *K"* ranged from 0.006-0.086 Pa.s$^{\pi}$ ' and 0.009-0.063 Pa.s$^{\pi}$ ", respectively. These values are lower than those reported by Augusto *et al.,* (2013b) in tomato juice, possibly due to the dissimilar working conditions (angular frequency range). The values of *n'and n* "were in the range of 1.546 to 2.124 and 1.351-1.942, respectively. These values are in agreement with those reported for tomato juice treated at high pressure (HPH). Since the magnitudes of G' were slightly higher than those of G", and the values of n'and n "were small but different from zero, the rheological behaviour of all nectars are described as weak gels. Similar results were reported by Moelants *et al.,* (2013) on carrot suspensions. Sharoba *et al.,* (2006) estimated under power law positive slope values in tomato juices and pastes, from which they declared the physical system of these products as weak gels. Likewise, Chen *et al.* (1996) inferred that globulin suspensions stabilised with corn starch and soy protein behave as gels, based on the estimation of viscoelastic parameters fitted to the power law model.

Comparing the values of the exponents estimated from the power law, we find that *n'* > *n"*, inferring that the elastic component is more frequency dependent than the viscous component (Falguera & Ibarz, 2014). This behaviour is clearly observed when beverages are subjected to high frequencies (>10rad/s). The effect of frequency and elastic character dominance on viscosity has been shown to be typical in the rheology of fruit juice suspensions and other vegetable-derived products (Ahmed *et al.,* 2007; Semero *et al.,* 2009; Augusto *et al.,* 2013b; Falguera & Ibarz, 2014). As the oscillatory frequency increases, the period of applied deformation decreases, and the time for structural rearrangement of the sample is too short. For this reason, the elastic deformations in the structure become more significant than the viscous properties (Ahmed *et al.,* 2007).

Table 3-7: Estimated viscoelastic parameters according to the power law model, in tree tomato beverages at a temperature of 30 C.°

Temperature (° C)

		10	20	30	40	50
K'	Control	0,029	0,036	0,039	0,021	0,075
	Minimum	0,011	0,006	0,008	0,013	0,027
	Central	0,035	0,034	0,033	0,026	0,046
	Maximum	0,043	0,014	0,019	0,013	0,027
K"	Control	0,029	0,009	0,032	0,011	0,050
	Minimum	0,012	0,006	0,010	0,017	0,032
	Central	0,036	0,042	0,053	0,014	0,028
	Maximum	0,049	0,017	0,016	0,009	0,010
n'	Control	1,831	1,809	1,795	1,546	1,652
	Minimum	1,942	2,084	1,983	1,945	1,762
	Central	1,710	1,719	1,718	1,78	1,577
	Maximum	1,620	1,875	1,769	1,942	1,771
n"	Control	1,461	1,769	1,468	1,762	1,357
	Minimum	1,636	1,784	1,623	1,564	1,351
	Central	1,466	1,429	1,287	1,641	1,485
	Maximum	1,391	1,629	1,589	1,659	1,709

Regression coefficients were greater than 0.95.

Fig. 4-4 shows the trend of the viscoelastic moduli as a function of temperature in tree tomato beverages. It can be seen that the moduli G' and G" decrease slightly with increasing temperature. However, there is a dissimilar behaviour of the modulus values when compared at high and low frequencies. For high angular frequencies (ω>10rad/s) no marked differences in the values of G' and G" are observed in the temperature range from 0 to 50°C. However, at low angular frequencies (ω<10 rad/s), the values of G' and G" tend to be slightly different with increasing temperature. This behaviour is probably due to the form of energy storage or loss, which occurs in the different frequency ranges. Chaikham and Apichartsrangkoon (2012) argue that energy storage at high frequencies is reversible, and is marked by elastic stretching of the molecular chains. However, at low frequencies the mode of energy storage and loss is dependent on the translational motion of the molecules, which is considered a measure of temperature.

However, the decrease in viscoelastic moduli with increasing temperature has been shown to be predominant in several fruit products. Sharoba *et al.,* (2006) report a decrease in G' and G" in tomato juices as temperature is increased in the range of O to 50 °C. Tiziani and Vodovotz, (2005) note a decrease in modulus in tomato juices with soy protein during heating up to 65°C. The same behaviour was reported in passion fruit pulp stabilised with guar gum and xanthan gum at low angular frequencies (<10 rad/s). To evaluate the effect of temperature on the viscoelastic behaviour of the beverages, the parameters *K', K", n' , n "were* fitted to the Arrhenius equation. The estimated models were not significant ($p>0.05$), and the coefficients of determination were relatively low. That is, no linear relationship was found for the moduli G' and G" versus temperature. Similar results were determined in tomato juices, when evaluating the oscillatory moduli from complex viscosity as a function of temperature (Tiziani and Vodovotz, 2005).

Fig. 4-5 shows the behaviour of the tangent or loss factor (δ) at different temperatures. It is observed that the values are below unity ($\tan \delta<1.0$) confirming the viscoelastic or solid gel behaviour of the formulated beverages (Rao, 2006). It was also found that the modulus ratio (G7G') is frequency dependent. It details a tendency for tan (δ) to remain constant at low frequencies (<1.0 Hz), but begins to decrease with increasing oscillation velocities (>1.0 Hz). A similar trend in the variation of tan (δ), in orange-milk juices stabilised with Persian gum was reported by Abbasi and Mohammadi (2013). Moraes *et al.* (2011) reported that tan (δ) in passion fruit pulp showed values less than unity with a weak frequency dependence for samples formulated with xanthan gum. However, the decrease in the G"/G' ratio at high frequencies confirms the predominance of the elastic component associated with polymeric materials (hydrocolloids) in the suspension, since these systems are close to true gels.

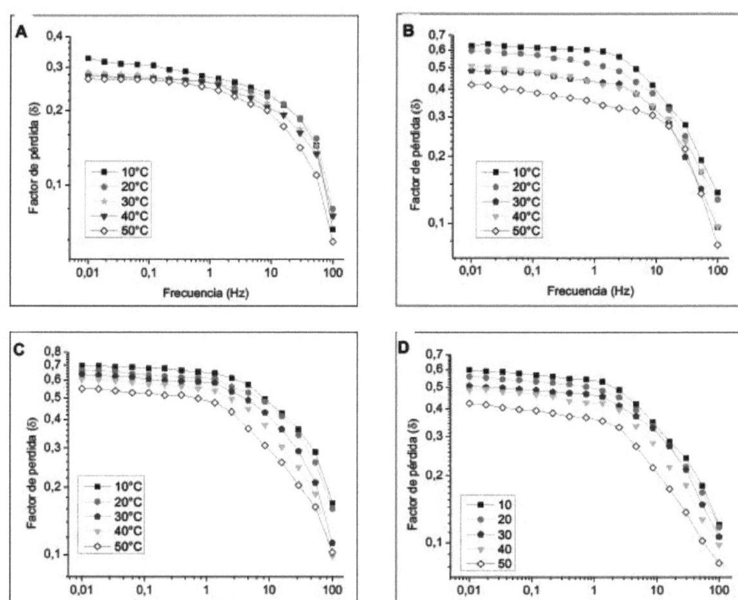

Figure 3-5: Behaviour of the tangent or loss factor (δ) as a function of temperature in tree tomato beverages

. A) Control, B) Minimum point, C) Midpoint, D) Maximum point.

Temperature exerted a slight influence on the decrease of the loss factor, however, no linear relationship was found from the Arrhenius equation. Sharoba *et al.*, (2006) in evaluating the viscoelastic properties of tomato juice and paste in the range of 0.01 to 100 Hz, found a decrease in tan (δ) values during heating. Likewise, Tiziani and Vodovotz, (2005) describe the effect of temperature on parameters such as complex and tangent viscosity, indicators of the viscoelastic behaviour of food dispersions. The authors describe that increasing temperature can affect the physical system of the juices that behave like a gel, causing slight damage to their structure, probably due to the weakening of hydrophilic and hydrophobic interactions between hydrocolloids and suspended tomato pulp particles. Similar characteristics have been reported by Choppe *et al.*, (2010) in xanthan gum suspensions subjected to frequency sweeps between 0.01 and 100 Hz. Lad and Murthy (2013) report that the increase in temperature caused a weakening in the cross-linked structure formed by the polysaccharide constituents of aloe vera gel, resulting in a decrease in tangent.

In order to evaluate the stability of the beverages under storage conditions, the behaviour of the viscoelastic parameters G', G" and tan (δ) was analysed at a temperature of 10° C and a frequency of 0.1 Hz. It was estimated that the addition of hydrocolloids and aloe vera did not influence the behaviour of the viscoelastic moduli ($p>0.05$), maintaining the trend G'>G" in the beverages. On the other hand, Falguera & Ibarz, (2014) suggest that at low frequencies, the physical system of the product is simulated under conditions of low deformation, as occurs during storage and in the sedimentation of suspended particles. The tan (δ) has been considered a relevant measure of physical stability in dispersions and it has been found that for tan (δ) values <1.0, dispersions are less susceptible to phase separation (Kuentz and Rothlisberger, 2003). Mezger (2006) argues that the weak gel structure (tan δ<1.0) is exhibited as a certain form of stability of food dispersions. Based on statistical analysis, the factors xanthan gum concentration and CMC concentration were found to be significant in the behaviour of tan (δ) in tree tomato beverages (p<0.05). The estimated mathematical model describing the behaviour of tan (δ) yielded a coefficient of determination of 0.89 (Table 4-8). Likewise, the model obtained in relation to the goodness-of-fit test showed a good statistical estimate (p>0.05).

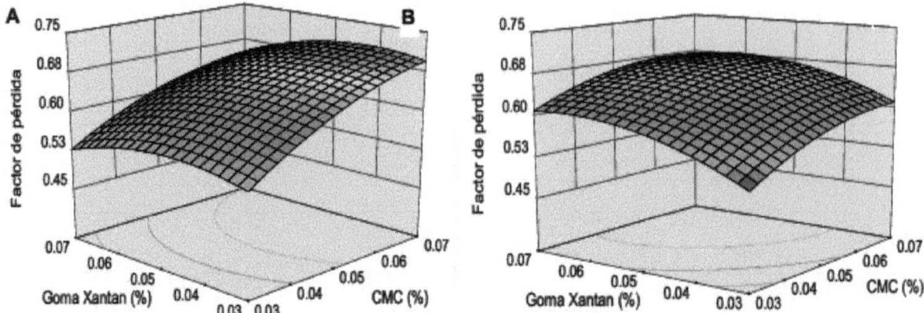

Figure 3-6: Response surface for loss tangent (tan δ) at 10° C and 0.1 Hz. A) 0.5% Aloe vera (minimum point), B) Aloe vera 1.5% (maximum point).

Fig. 4-6 shows that as the hydrocolloid concentration increases, the tan δ values increase, but the weak gel structure is preserved (tan δ<0.1). Similar results were reported by Moraes *et al.* (2011) in passion fruit juices treated with xanthan gum. Authors cite that these characteristics can be explained by the fact that hydrocolloids in suspensions tend to form a highly ordered complex structure in the form of a network of rigid molecules, particularly a weak gel, contributing to the stability of particles in suspension (Rao, 2006; Norton *et al.*, 2011). This is in agreement with Meng and Rao (2005) who argue that the physical gel-like behaviour of some food suspensions is mainly attributed to the presence of polymeric substances. Chaikham and Apichartsrangkoon (2012) found a predominance of the elastic component in *Dimocarpus longan* juices using 0.1% xanthan concentration, for which tangent values of less than unity were estimated. The increase in viscoelastic component values in apple juice suspensions is derived from the use of cross-linked corn starches (Meng and Rao, 2005). They point out that increasing the concentration of starch in the suspension generates an increase in the values of the tangent at low frequencies (<10Hz).

Table 3-8: Regression model effects for loss tangent (δ) at 10° C and 0.1 Hz.

Regression coefficients	tan (δ)
βo	0,43
βl	-0,19
$\beta 2$	1,77
β_3	8,72
$\beta 11$	NS
$\beta l2$	3,80
$\beta\ 13$	-2,22
$\beta 22$	-47,21
$\beta\ 3_2$	-11,20
$\beta\ 3_3$	-37,57
R^2	0,89
Model (p-value)	0,04
Missing fit (p-value)	0,96

*NS: Coefficients not significant

A maximum concentration of 0.15% xanthan has been found to confer optimal viscoelastic characteristics in a fruit juice, conferring stability based on the dominance of the elastic component (G'>G") and a tan (δ) value <1.0 (Chaikham and Apichartsrangkoon, 2012).

Benchabane and Bekkour (2008) report that low concentrated CMC suspensions (<2.0%) can confer good stability due to their weak gel-like behaviour. From the above, desirable viscoelastic properties (G'>G") with tan (δ) values <1.0, in tree tomato beverages are perceived for xanthan gum concentrations greater than or equal to 0.025%, and CMC concentrations greater than or equal to 0.05%. Moreover, under these same concentrations, a pseudoplastic behaviour (*n<1*) and a significant increase in the viscosity of the continuous phase (increase of the consistency coefficient, *K*, parameters that are related as indicators of good stability in suspensions) are perceived. Tiziani and Vodovotz 2005) argue that the high water-holding capacity and ability to increase the viscosity of the continuous phase of hydrophilic colloids, highlight their particle retention and ionisation properties, forming in turn a stronger gel-like structure (G'>G") conferring stability to food suspensions. On the other hand, it is observed that xanthan gum can exert better stabilising properties than CMC gum at lower concentrations. Some authors point out that despite its high cost, xanthan gum remains the first choice due to its unparalleled rheological properties. They state that the ability to confer higher viscosity at low shear rates (<10s"^{1}), increased viscoelastic properties and lower creep deformation account for its excellent stabilising properties (Lazaridou *et al.*, 2007; Li *et al.*, 2015).

CONCLUSIONS

The present study has demonstrated the significant effect of hydrocolloid addition on the rheological properties of tree tomato beverages. The flow characteristics of all beverages were best described by the Power Law model with a high correlation coefficient ($R^2 \approx 1$). The decrease of the flow behaviour index *(n)* marked by the incorporation of polymeric materials indicates the pseudoplastic behaviour of the formulated beverages. The Arrhenius principle was adequately applied to describe the effect of temperature on the consistency coefficient and flow behaviour index. The viscosity of the formulated nectars decreased significantly with increasing temperature.

Dynamic oscillatory tests confirm the weak gel behaviour of the products given values of *n'* and *n"* greater than unity, and predominance of the elastic modulus (G'>G"). The addition of hydrocolloids and aloe increased the values of the moduli G' and G". The increase in temperature did not show a linear trend in the parameters describing the viscoelastic behaviour, based on the Arrhenius principle. The consistency coefficient *(K)* and loss tangent (tan δ) were found to be significant with the addition of hydrocolloids (p>0.05). Concentrations of xanthan gum ≥ 0.025% and CMC ≥ 0.05% can confer good physical stability, thus decreasing the phenomenon of sedimentation and phase separation in tree tomato drinks.

ACKNOWLEDGEMENTS

The authors would like to thank the Directorate of Research (DIME) of the National University of Colombia - Medellin, for the financial support of the research project registered under number 19822, through the call "National Programme of Projects for the strengthening of research, creation and innovation in postgraduate studies 2013-2015".

BIBLIOGRAPHY

Abbasi, S., Mohammadi, S. (2013). Stabilization of milk-orange juice mixture using persian gum: Efficiency and mechanism. *Food Bioscience,* 2(1), 53-60.

Ahmed, J., Ramaswamy, H. S., Sashidhar, K. C. (2007). Rheological characteristics of tamarind *(Tamarindus indica L.)* juice concentrates. *LWT - Food Science and Technology,* 40(2), 225-231.

Augusto, P. E., Ibarz, A., Cristianini, M. (2013a). Effect of high-pressure homogenization (HPH) on the rheological properties Oftomatojuice: Viscoelastic properties and the Cox-Merz rule. *Journal of FoodEngineering,* 114(1), 57-63.

Augusto, P., Ibarz, A., Cristianini, M. (2013b). Effect of high-pressure homogenization (HPH) on the rheological properties Oftomatojuice: Creep and recovery behaviours. *Food Research International, 54(1), 169-176.*

Banerjee, I., Ghosh, U. (2015). Effect Ofdifferent hydrocolloids on rheology Oftamarind *(Tamarindus indica* L.) juice. *InternationalJournalofLatest Trendsin Engineering and Technology, 5(1),* 60-70.
Benchabane, A., Bekkour, K. (2008). Rheological properties of carboxymethyl cellulose (CMC) solutions. *Colloid and PolymerScience,* 286(10):1173-1180.
Boghani AH, Raheem A, Hashmi SI (2012) Development and storage studies of blended papayaaloe vera ready to serve (RTS) beverage. *Journal of Food Processing & Technology,* 3(19), 185189.
Coffey, D. G., Bel, D. A., Henderson, A. Cellulose and cellulose derivatives. In: Stephen, A. M., Phillips, G. O., Williams, P. A. Food polysaccharides and their applications. Boca Raton, USA: CRC Press, 2006.
Cabral, R. A. F., Orrego, A. C. E., Gabas, A. L. Telıs, R. J. (2007). Rheological and thermophysical properties of blackberry juice. *Ciência e Tecnologia de Alimentos - Campinas,* 27(3): 589-596.
Chaikham. P., Apichartsrangkoon, A. (2012). Comparison of dynamic viscoelastic and physicochemical properties of pressurised and pasteurised longan juices with xanthan addition. *Food Chemistry,* 134 (4), 2194-2200.
Chen, C. J., Liao, H. J.; Okechukwu, P. E.; Damodaran, S., Rao, M. A. (1996). Rheological properties of heated corn starch + soybean 7S and 11S globulin dispersions. *Journal of Texture Studies,* 27(4), 419-432.
Chin, N. L., Chan, S. M., Yusof, Y. A., Chuah, T. G., Talıb, R. A. (2009). Modelling of rheological behaviour of pummelo juice concentrates using master-curve. *Journal of Food Engineering,* 93(2), 134-140.
Choi, S. and Chung, M. (2003). A review on the relationship between Aloe vera components and their biologic effects. *Seminars in Integrative Medicine,* 1,53-62.
Choppe, E., Puaud, F., Nicolai, T., Benyahia, L. (2010). Rheology Ofxanthan solutions as a function Oftemperature, concentration and ionic strength. *Carbohydrate Polymers,* 82 (4), 1228-1235.
Dak, M., Verma, R.C., Sharma, G.P. (2006). Flow characteristics of juice of Totapuri mangoes. *JournalofFood Engineering,* 76(4), 557-561.
Dominguez, F. R., Arzate, V. I., Chanona, P. J., Welti, C. J., Alvarado, G. J., calderón, D. G., Garibay, F. V., Gutiérrez, L. G. (2012). Aloe vera gel: structure, composition, chemistry, processing, biological activity and importance in the pharmaceutical and food industry. *Revista Mexicana de Ingeniería Química,* 11(1),23-43.
Earle, R. L. Unit Operations in Food Processing. Second Edition. London, UK: Pergamon Press, 1989, 207 p.
Falcone, P.M., Chillo, S., Giudici, P., Del Nobile, M.A. (2007). Measuring rheological properties for applications in quality assessment of traditional balsamic vinegar: Description and preliminary evaluation of a model. *Journal of Food Engineering,* 80 (1), 234-240.
Falguera, V., Ibarz, A. Juice Processing : Quality, Safety and Value-Added Opportunities. First Edition. Boca Raton, Florida: CRC Press, 2014, 401 p.
Fasolin, L., and Cunha, R. (2012). Soursop juice stabilized with soy fractions: A rheologıal approach. *Ciência e Tecnologia de Alimentos,* 32(2), 558-567.
García, O. F., Santos, V. E., Casas, J. A., Gómez, E. (2000). Xanthan gum: production, recovery, and properties. *BiotechnologyAdvances,* 18(7), 549 - 579.
Genovese, D. B., Elustondo, M. P., Lozano, J. E. (1997). Colour and cloud stability in cloudy apple juice by steam heating during crushing. *JournalofFoodScience,* 62(6), 1171-1175.
Genovese, D. B., Lozano, J. E. (2001). The effect of hydrocolloids on the stability and viscosity of cloudy apple juices. *FoodHydrocolloids,* 15(1), 1-7.
Genovese, D., Lozano, J. (2006). Contribution of colloidal forces to the viscosity and stability of cloudy apple juice. *Food Hydrocolloids,* 20(6), 67-773.
Goula, A. M., Adamopoulos, K. G. (2011). Rheological models of kiwifruit juice for processing applications. *FoodProcessing & Technology,* 2(1), 106-113.
Gratão, A.C, Silveira Jr., V., Telis-Romero, J. (2007). Laminar flow of soursop juice through concentric annuli: friction factors and rheology. *JournalofFood Engineering 78 (4),* 1343-1354.
Habeeb, F., Shakir, E., Bradbury, F., Cameron, P., Taravati, M. R., Drummond, A. J., Gray, A. I., Ferro, V. A. (2007). Screening methods used to determine the anti-microbial properties of Aloe vera innergel. *Methods,* 42, 315-320.
Haminuik, C. W., Sierakowski, M. R., Vidal, J. M., Masson, M. L. (2006). Influence Oftemperature on

the rheological behaviour of whole araca pulp *(Psidium Cattleianum sabine). Food Science and Technology,* 39(4), 426-430.
Hamman, J.H. (2008). Composition and applications of Aloe vera leaf gel. Molecules, 13(8), 15991616.
Ibrahim, G., Hassan, I., Abd-Elrashid, A., El-Massry, K., Eh-Ghorab, A., Ramadan, M., Osman, F. (2011). Effect of clouding agents on the quality of apple juice during storage. *Food Hydrocolloids,* 25(1), 91-97.
Colombian Institute of Technical Standards. NTC 3549. Fruit soft drinks. Bogotá: ICONTEC, 1999.
Colombian Institute of Technical Standards. NTC 4105. Fresh fruits. Tree tomato, Specifications. Fruit drinks. Bogotá: ICONTEC, 1997.
Kou, M., Yen, J., Hong, J.; Wang, C.; Lin, C.; Wu, M. (2009). *Cyphomandra betacea Sendt.* phenolics protect LDL from oxidation and PC12 cells from oxidative stress. *LWT - Food Science and Technology,* 42, 458-463.
Krokida, M.K., Maroulis, Z.B., Saravacos, G.D., 2001. Rheological properties of fluid fruit and vegetables puree products: Compilation Ofliterature data. *International Journal of Food Properties,* 4 (2), 179-200.
Kuentz, M., Rothlisberger, D. (2003). Rapid assessment of sedimentation stability in dispersions using near infrared transmission measurements during centrifugation and oscillatory rheology. *European Journal OfPharmaceutics and Biopharmaceutics,* 56(3), 355-361.
Kumar, S., Kumar, P. (2015). Rheological modeling of non-depectinized beetroot juice concentrates. *JournalofFoodMeasurementandCharacterization, 1, 1-8.*
Lad, V. N., Murthy, Z.V.P. Rheology of Aloe barbadensıs Miller: A naturally available material of high therapeutic and nutrient value for food applications. *Journal of Food Engineering,* 115(3), 279-284.
Lazaridou, A., Duta, D., Papageorgiou, M., Belc, N., Biliaderis, C. G. (2007). Effects of hydrocolloids on dough rheology and bread quality parameters in gluten-free formulations. *Journal of Food Engineering,* 79(3), 1033-1047.
Liang, C., Hu., X., Ni, Y., Wu., J., Chen, F., Liao, X. (2006). Effect of hydrocolloids on pulp sediment, white sediment, turbidity and viscosity of reconstituted carrot juice. *Food Hydrocolloids,* 20(8), 11901197.
Li, J. M., Nie, S. P. (2015). The functional and nutritional aspects of hydrocollοιds in foods. *Food Hydrocolloids,* In Press, 1-16.
Márquez, C., Otero, C., Cortes, M. (2007). Physiological, textural, physicochemical and microstructural changes of tree tomato *(Cyphomandra betacea S.)* in postharvest. *Journal of the Faculty of Pharmaceutical Chemistry,* 14 (2), 9-16.
Meng, Y., Rao, M. (2005). Rheological and structural properties of cold-water-swelling and heated cross-linked waxy maize starch dispersions prepared in apple juice and water. *Carbohydrate Polymers,* 60(3), 291-300.
Meza, N., Manzano, M. J. (2009). Tree tomato *(Cyphomandra betaceae [Cav.] Sendtn)* fruit characteristics based on aril coloration in the Venezuelan Andean Zone. *Revista UDO Agricola,* 9(2), 289-294.
Mezger, T. G. (2006). The rheology handbook. Second Edition. Hannover: Vincentz Network, 300p.
Moelants, K. R. N., Cardinaels, R., Jolie, R. P., Verrijssen, T. A. J., Van Buggenhout, S., Zumalacarregui, L. M., Van Loey, A. M., Moldenaers, P., Hendrickx, M. E. (2013). Relation between particle properties and rheological characteristics of carrot-derived suspensions. *Food Bioprocess Technology,* 6(5):1127-1143.
Moraes, I. C. F. F., Fasolin, L. H., Cunha, R. L., Menegalli, F. C. (2011). Dynamic and steady-shear rheological properties of xanthan and guar gums dispersed in yellow passion fruit pulp *(Passiflora edulis flavicarpa). Brazilian Journal of Chemical Engineering,* 28(3), 483 - 494.
Norton, L., Spyropoulos, F., Cox, P. Practical Food Rheology-An Interpretive Approach. First Edition. Oxford, UK: Wiley-Blackwell, 2011,280 p.
Paquet, E., Bedard, A., Lemieux, S., Turgeon, S. L. (2014). Effects of apple juice-based beverages enriched with dietary fibres and xanthan gum on the glycemic response and appetite sensations in healthy men. *Bioactive Carbohydrates and DietaryFibre,* 4(1), 39-47.
Quek, M. C, Chin, N. L., Yusof, Y. A (2013). Modelling of rheological behaviour of soursop juice concentrates using shear rate-temperature-concentration superposition. *Journal of Food Engineering,*

118(4), 380-386.
Ramachandra, C.T., and Srinivasa, P. (2008). Processing of Aloe vera gel: A review. *American Journal OfAgricultural and Biological Sciences,* 3(2), 502-510.
Rao, M. A. Rheology of Fluid and Semisolid Foods: Principles and Applications. Second Edition. Geneva, NY, USA: Springer, 2006, 481p.
Sesmero, R., Mitchell, J. R., Mercado, J. A., Quesada, M. A. (2009). Rheological characterisation of juices obtained from transgenic pectate lyase-Silenced strawberry fruits. *Food Chemistry,* 116(2), 426-432.
Shamsudin, R., Chia, S. L., Mohd, A. N., Wan, D. W. (2013). Rheological properties of ultraviolet-irradiated and thermally pasteurized Yankee pineapple juice. *Journal of Food Engineering,* 116(2), 548-553.
Sharoba, A. M., Senge, B., El-Mansy, H. A., Bahlol, H. E. M., Blochwitz, R. Rheological properties of some Egyptian and European tomato products. In: First Inter. Conf. & Exh. Food & Tourism. Egypt: HITH, 2006, p. 316-334.
Sogi, D. S., Oberoi, D. P., Malik, S. (2010). Effect of particle size, temperature, and total soluble solids on the rheological properties ofwatermelon juice: A response surface approach. *International JournalofFoodProperties,* 13(6), 1207-1214.
Tızıanı, S., Vodovotz, Y. (2005). Rheological effects of soy protein addition to tomato juice. *Food Hydrocolloids,* 19 (1), 45-52.
Vandresen, S., Quadri, M. G., de Souza, J. A., Hotza, D. (2009). Temperature effect on the rheological behaviorof carrot juices. *JournalofFoodEngineering*, 92(3), 269-274.
Vasco, C., Avila, J., Rúales, J., Svanberg, U. and Kamal-Eldin, A. (2009). Physical and chemical characteristics of golden-yellow and purple-red varieties of tamarillo fruit *(Solanum betaceum Cav.). InternationaljournalofFood Sciences and Nutrition*, 60(S7), 278-288.
Xu, L., Xu, G., Liu, T., Chen, Y., Gong, H. (2013). The comparison of rheological properties of aqueous welan gum and xanthan gum solutions. *Carbohydrate Polymers,* 92 (1), 516- 522.
Yogurtçu, H., Kamişli, F. (2006). Determination of rheological properties of some pekmez samples in Turkey. *JournalofFood Engineering,* 77(4), 1064-1068.
Whistler, R. L., BeMiller, J. N. Carbohydrate Chemistryfor Food Scientists. Second Edition. St. Paul MN, USA: Eagan Press, 1997.
Williams, P.A. and Phillips, G.O. Introduction to food hydrocolloids. In Phillips, G.O., Williams, P.A. Handbook of hydrocolloids. Cambridge, United Kingdom: CRC Press, 2009.
Nwaokoro. O, G., Akanbi, C. T. (2015). Effect of the addition of hydrocollοids to tomato-carrot juice blend. *JournalofNutritionalHealth & Food Science,* 3(1): 1-10.

Conclusions and recommendations

The present study has demonstrated the significant effect of the addition of hydrocolloids on the rheological properties and degree of stability in tree tomato beverages. Concentrations of xanthan gum and CMC, equal or higher than 0.05%, can be considered as suitable treatments in the control of the related physical instability at low settling velocities and high zeta potential values (>30 mV), without significantly affecting physicochemical properties, colour parameters and sensory attributes. The incorporation of aloe vera gel significantly affected pH and titratable acidity values, however, it did not influence the instability of the beverages.

The results suggest the occurrence of two physical stability mechanisms in tree tomato drinks: spherical and electrostatic. The first mechanism is related to the hydrophilic capacity of xanthan gum to absorb and retain moisture, producing an increase in the viscosity of the continuous phase, helping to retain particles in suspension, and increasing the turbidity of the suspension. The second is associated with the anionic character of the hydrocolloid to exert a significant effect on the repulsive forces between particles, considering the behaviour of the Z potential.

Flow properties describe the pseudoplastic behaviour of the beverages ($n<1.0$). A significant effect of xanthan gum addition on rheological parameters such as consistency coefficient (*K*) and flow index (*n*), estimated from the power law with excellent statistical fit, was estimated. Based on the Arrhenius principle, it was possible to evaluate the incidence of temperature on theological parameters, observing a decrease in viscosity reflected in the reduction of consistency coefficients with increasing temperature. Tree tomato nectars have theological characteristics of weak gel, based on values of n' and n'' greater than unity, and predominance of elastic modulus ($G'>G''$). The moduli G' and G" are influenced by the gradual increase in temperature, however, the parameters describing the viscoelastic behaviour of the beverages did not conform to the Arrhenius principle.

The addition of xanthan gum significantly affected the loss factor. Since at concentrations of xanthan gum ≥0.05% and CMC≥ 0.05% the weak gel behaviour and $\delta<1.0$ are maintained, these treatments can be classified as adequate in the evaluation of the degree of physical stability of the beverages. Moreover, the inclusion of gums under these concentrations contributes to control the phenomenon of sedimentation and phase separation in the beverages, preserving the intrinsic physicochemical and sensory characteristics of tree tomato pulp.

Bibliography

Abbasi, S., Mohammadi, S. (2013). Stabilization of milk-orange juice mixture using Persian gum: Efficiency and mechanism. *FoodBioscience,* 2, 53-60.

Ahlawat, K. S., Khatkar, B. S. (2011). Processing, food applications and safety of aloe vera products: A review. *Journal ofFood Science and Technology,* 48(5), 525-533.

Alarcon-Restrepo, J. J., Chavarriaga-Montoya, W. (2007). Early diagnosis of anthracnose *(Colletotrichum gloeosporioides)* (Penz) Penz & Sacc. on tree tomato using quiescent infections. *Agronomy,* 15(1): 89 - 102.

Alves, D.S., Perez, F. L., Estepa, A., Micol, V. (2004). Membrane-related effects underlying the biological activity of the anthraquinones emodin and barbaloin. *Biochemical Pharmacology,* 68, 549-561.

Amaya, J., Hashimoto, J. Tree tomato *(Cyphomandra betacea* Sendt). Gerencia Regional de Recursos Naturales y Gestión del Medio Ambiente. Trujillo: GRNM, 2006, 8p.

Aranberri, I., Binks, B., Clint, J., Fletcher, P. (2006). Development and characterisation of emulsions stabilised by polymers and surfactants. *Revista Iberoamericana de Polímeros,* 7(3), 211-231.

Arslan, E. Rheological characterization of tahın/pekmez (sesame paste/concentrated grape juice) blends. Master in Science. Turkey, 2003, 59 p. Department of Food Engineering. Middle East Technical University.

Augusto, P., Ibarz, A., Cristianini, M. (2013). Effect of high-pressure homogenization (HPH) on the rheological properties of tomato juice: Creep and recovery behaviours. *Food Research International, 54(1), 169-176.*

Benitez, E. I., Genovese, D. B., Lozano, Jorge. E. (2009). Effect oftypical sugars on the viscosity and colloidal stability ofapplejuice. *FoodHydrocolloids,* 23, 519- 525.

Benítez, S., Achaerandio, I., Sepulcre, F., Pujóla, M. (2013). Aloe vera based edible coatings improve the quality of minimally processed 'Hayward' kiwifruit. *Postharvest Biology and Technology,* 81, 29-36.

Black, N., Ortega, L. Use of modified atmospheres in the preservation of babaco, tree tomato and passion fruit. Salgolqui, 2005, 187 p. Degree work (Agricultural Engineer). Army Polytechnic School. Faculty of Agricultural Sciences.

Boghani, A. H., Raheem, A., Hashmi, S. I. (2012). Development and storage studies of blended papaya-aloe vera ready to serve (RTS) beverage. *Journal of Food Processing & Technology,* 3, 185.

Boudreau, M. and Beland, F. (2006). An evaluation of the biological and toxicological properties of *Aloe barbadensis* Miller (Aloe vera). *Journal of Environmental Science and Health, Part C 24,* 103-154.

Brito, B., Espin, S., Villacres, E., Vaillant, F., Torres, N. and Sanaicela, D. Tree tomato. Physical and nutritional characteristics of the fruit, important in the research and development of pulps and chips. Instituto Nacional Autónomo de Investigaciones Agropecuarias. Quito: INIAP, 2008, 2 p.

Cáceres-Miranda, Lorena. Postharvest handling of tree tomato fruits *(Cyphomandra betacea)* and its relationship with shelf life in the central market of Cantón (Ambato). Ambato, 2012, 240p. Degree thesis (Master in Cleaner Production). Technical University of Ambato. Faculty of Food Science and Engineering.

Calvo-Villegas, Iván. Tree tomato *(Cyphomandra betacea)* cultivation. Integrated crop management. Instituto Nacional de Innovación y Transferencia en Tecnología Agropecuaria. Costa Rica: INTA, 2009, 6 p.

Castro, H., Benelli, P., Ferreira, S., Parada, F. (2013). Supercritical fluid extracts from tamarillo *(Solanum betaceum* Sendtn) epicarp and its application as protectors against lipid

oxidation ofcooked beefmea. *JournalofsupercriticalFluids,* 76, 17-23.
Castro-Parra, Andrea. Effect of the application of edible coatings on postharvest quality of tree tomato *(Solanum betaceum).* Quito, 2013, 138p. Degree thesis (Agroindustrial Engineer). National Polytechnic School. Faculty of Chemical Engineering and Agroindustry.
Chaikham, P., Apichartsrangkoon, A. (2012). Comparison of dynamic viscoelastic and physicochemical properties of pressurised and pasteurised longan juices with xanthan addition. *Food Chemistry,* 134(4), 2194-2200.
Chandan, B., Saxena, A. K., Shukla, S., Sharma, N., Gupta, D., Suri, K., Suri, J., Bhadauria, M., Singh, B. (2007). Hepatoprotective potential of *Aloe barbadensis* Miller. Against carbon tetrachloride induced hepatotoxicity. *JournalEthnopharmacoly,* 111, 560-566.
Chin, N. L., Chan, S. M., Yusof, Y. A., Chuah, T. G., Talıb, R. A. (2009). Modelling of rheological behaviour of pummelo juice concentrates using master-curve. *Journal of Food Engineering,* 93(2), 134-140.
Chivero, P., Gohtani, S., Yoshii, H., Nakamura, A. (2015). Effect Ofxanthan and guar gums on the formation and stability of soy soluble polysaccharide oil-in-water emulsions. *Food Research International*, 70, 7-14.
Choi, S., Chung, M. (2003). A review on the relationship between Aloe vera components and their biologic effects. *Seminars in Integrative Medicine,* 1, 53-62.
Chun-Hui, L., Chang-Hai, W., Zhi-Liang, X., Yi, W. (2007). Isolation, chemical characterization and antioxidant activities oftwo polysaccharides from the gel and the skin of *Aloe barbadensis* Miller irrigated with sea water. *Process Biochemistry,* 42, 961-970.
Croak, S., Corredig, M. (2006). The role of pectin in orange juice stabilization: Effect of pectin methylesterase and pectinase activity on the size of cloud particles. *Food Hydrocolloids,* 20, 961-965.
Davis, K., Philpott, S., Kumar, D., Mendall, M. (2006). Randomised double-blind Placebocontrolled trial of *Aloe vera* for irritable bowel syndrome. *International Journal of ClinicalPractice,* 60, 1080-1086.
Dominguez, F. R., Arzate, V. I., Chanona, P. J., Welti, C. J., Alvarado, G. J., Calderon, D. G., Garibay, F. V., Gutiérrez, L. G. (2012). Aloe vera gel: structure, composition, chemistry, processing, biological activity and importance in the pharmaceutical and food industry. *Revista Mexicana de Ingeniería Química,* 11(1),23- 43.
Do-Sang L, Ryu II, Kap-Sang L, Yang-See S, Seung-Ho C (1999). Optimisation in the preparation of aloe vinegar by *Acetobactor sp.* and inhibitory effect against lipase activity. *JournaloftheKoreanSocietyforAppliedBiological Chemistry,* 42(2), 105-110.
Drago, S. M., Lopez, L. M., Sainz, E. T. (2006). Bioactive components of functional foods of plant origin. *Revista Mexicana de Ciencias Farmacéuticas,* 37(4), 58-68.
Duque, A., Giraldo, G., Quintero, V. (2011). Characterization of cape gooseberry *(Physalisperuviana L.)* fruit, pulp and concentrate. *Temas agrarios,* 16(1), 75-83.
Elbandy, M. A., Abed, S. M., Gad, S. A., Abdel-Fadeel, M. G. (2014). Aloe vera gel as a functional ingredient and natural preservative in mango nectar. *World Journal of Dairy & Food Sciences,* 9 (2), 191 - 203.
Eshun, K., He, Q. (2004) Aloe vera: A valuable ingredient for the food, pharmaceutical and cosmetic industries. *Critical Reviews in Food Science and Nutrition,* 44 (2), 91-96.
Farfan, D., Zambrano, Mayra. Business plan to market tree tomatoes from Ecuador to the United States of America. Ecuador, 2010, 94 p. Degree work (Engineer in Foreign Trade and International Business). Universidad Laica Eloy Alfaro de Manabi. Faculty of Commerce and International Business.
Fasolin, L., Cunha, R. (2012). Soursop juice stabilized with soy fractions: A rheologıal approach. *Ciência e Tecnologia deAlimentos,* 32(2), 558-567.
Fayyaz, A., Sani, M. A, Arian, A. F. (2014). The effect of some hydrocolloids on inhabitation

ofserum separation ofdoogh. *DAMA International,* 3(2), 259 - 266.
Ferrer, L. B., Dalmau, S. J. (2001). Functional foods: probiotics. *Acta Pediatrica Española,* 59, 150-155.
García, M., García, H. Harvest and post-harvest management of blackberry, lulo and tree tomato. Corporación Colombiana de Investigación Agropecuaria. Corpoica: Bogotá, 2001, 102 p. García-Muñoz, María. Manual de manejo cosecha y postcosecha del tomate de árbol. Corporación Colombiana de Investigación Agropecuaria. Bogotá: Corpoica, 2008, 98 p.
Garriga, A. M. Rheology of cellulosic thickeners for waterborne paints: Modelling and associative thickening mechanism. Barcelona, 2002, 76 p. Doctoral thesis. University of Barcelona. Department of Chemical Engineering.
Genovese, D., Elustondo, M., Lozano, J. (1997). Colour and cloud stabilization in cloudy apple juice by steam heating during crushing. *JournalofFood Science,* 62, 1171-1175.
Genovese, D., Lozano, J., Rao, M. (2007). The rheology ofcolloidal and noncolloidal food dispersions. *JournalofFoodScience,* 72(2), 11-20.
Genovese, D., Lozano, J. (2001). The effect of hydrocolloids on the stability and viscosity Ofcloudyapplejuices. *FoodHydrocolloids,* 15, 1-7.
Giupponi, G., and Pagonabarraga, I. (2011). Determination of the zeta potential for highly charged colloidal suspensions. *Philosophical Transactions of the Royal Society A,* 369, 2546-2554.
Goula, A. M., Adamopoulos, K. G. (2011). Rheological models of kıwıfruıt juice for processing applications. *FoodProcessing & Technology,* 2(1), 106-113.
Gratão, A, Silveira Jr., Telis, R. J. (2007). Laminarflow Ofsoursopjuice through concentric annuli: friction factors and rheology. *JournalofFoodEngineering 78 (4),* 1343-1354.
Guerrero, S. N., Alzamora, S. M. (1998). Effect of pH, temperature and glucose addition on flow behavior of fruit purees: II. Peach, papaya, and mango purees. *Journal of Food Engineering,* 33, 239-256.
Habeeb, F., Shakir, E., Bradbury, F., Cameron, P., Taravati, M. R., Drummond, A. J., Gray, A. I., Ferro, V. A. (2007). Screening methods used to determine the anti-microbial properties of *Aloe vera* inner gel. *Methods,* 42, 315-320.
Hamman, J. H. (2008). Composition and applications of *Aloe vera* leaf gel. *Molecules,* 13, 1599-1616.
Hu, Q., Hu, Y., Xu, J. (2005). Free radıcal-scavenıng activity of Aloe vera *(Aloe barbadensıs* Miller) extracts by supercritical carbon dioxide extraction. *Food Chemistry,* 91, 85-90.
Ibrahim, G., Hassan, I., Abd-Elrashid, A., El-Massry, K., Eh-Ghorab, A., Ramadan, M., Osman, F. (2011). Effect of clouding agents on the quality of apple juice during storage. *FoodHydrocolloids,* 25(1), 91-97.
Im, S. A., Oh, S. T., Song, S., Kim, M. R., Woo, S. S., Jo, T. H., Park, Y. I., Lee, C. K. (2005). Identification of optimal molecular size of modified Aloe polysaccharides with maximum immunomodulatory activity. *Internationalallmmunopharmacology,* 5, 271-279.
Jayme, M. L., Dunstan, D. E., Gee, M. L. (2009). Zeta potential of gum arabic stabilized oillnwateremulsions. *FoodHydrocolloids,* 13(6), 459-465.
Jibaja-Mera, Hugo. Modelling of oil absorption kinetics during the vacuum frying process of tree tomato *(Solanum betaceum.)* flakes. Quito, 2010, 148p. Degree thesis (Agroindustrial Engineer). National Polytechnic School. Faculty of Engineering and Chemistry.
Juárez, I., M. Present and future of functional foods. In: Juárez, I., M., Perote, A., A. Healthy and specifically designed foods: Functional foods. Madrid, Spain: IMCS.A.,2010. p.30-45.
Kaneiwa, M., Augusto, P., Cristianini, M. (2013). Effect of high-pressure homogenization (HPH) on the physical stability Oftomatojuice. *FoodResearch International,* 51, 170-179.
Kaya, A., Sözer, N., (2005). Rheological behaviour of sour pomegranate juice concentrates *(Punica granatum). InternationaljournalofFood Science & Technology,* 40 (2), 223-227.

Keshtkaran, M., Mohammadifar, M. A., Asadi, G. A., Nejad, R. A., Balaghi, S. (2013). Effect ofgum tragacanth on rheological and physical properties of a flavored milk drink made with date syrup. *JournalofDairy Science,* 96(8), 4794-4803.

Khoshgozaran, A. S., Hossein, A. M., Hamidy, Z., Bagheripoor, F. N. (2012). Mechanical, physicochemical and colour properties of chitosan based-films as a function of aloe vera gel incorporation. *CarbohydratePolymers,* 87, 2058-2062.

Kim, K., Kim, H., Kwon, J., Lee, S., Kong, H., Im, S., Lee, Y., Oh, S., Jo, T., ParkY., Lee, C., Kim K. (2009). Hypoglycemic and hypolipidemic effects of processed *Aloe vera* gel in a mouse model ofnon-insulin-dependent diabetes mellitus. *Phytomedicine,* 16, 856-863.

Kou, M., Yen, J., Hong, J., Wang, C., Lin, C., Wu, M. (2009). *Cyphomandra betacea* Sendt. phenolics protect LDL from oxidation and PC12 cells from oxidative stress. *LWT - Food Science and Technology,* 42, 458-463.

Kumar, S., Kumar, P. (2015). Rheological modeling of non-depectınızed beetroot juice concentrates. *JournalofFoodMeasurementandCharacterization, 1, 1-8.*

Lagos-Santander, Liz. Evaluation of the genetic potential of some tree tomato *(Cyphomandra betacea Cav.* Sendt) parents from a circulating partial dialelic. Palmira, 2012, 79 p. Degree work (Master in Agricultural Sciences). National University of Colombia. Faculty of Agricultural Sciences.

Langmead, L., Makins, R. J., Rampton, D. S. (2004). Anti-inflammatory effects of Aloe vera gel in human colorectal mucosa in vitro. *Alimentary Pharmacology and Therapeutics,* 19, 521-527.

Lazaridou, A., Duta, D., Papageorgiou, M., Belc, N., & Biliaderis, C. G. (2007). Effects of hydrocolloids on dough rheology and bread quality parameters in gluten-free formulations. *JournalofFoodEngineering*, 79(3), 1033-1047.

Lee, J., Hand, Y. H (1997). Characteristics of aloe vera suspended liquid yoghurt inoculated with *Lactobacillus Casei* YIT 9018. *Korean JournalofAnimal Science,* 39, 93-100.

Lee, M., Lee, O. Yoon, S. (1998). In vitro angiogenic activity of aloe vera gel on calf pulmonary artery endothelial (CAPE) cells. *Archives Pharmacal Research* 21, 260-265.

León, J., Viteri, P., Cevallos, G. Manual de cultivo de tomate de árbol. Instituto Nacional Autónomo de Investigaciones Agropecuarias (INIAP). Ecuador: INIAP, 2004, 70p.

Li, J. M., Nie, S. P. (2015). Thefunctional and nutritional aspects Ofhydrocolloids in foods. *FoodHydrocolloids,* In Press, 1-16.

Liang, C., Hu., X., Ni, Y., Wu., J., Chen, F., Liao, X. (2006). Effect Ofhydrocolloids on pulp sediment, white sediment, turbidity and viscosity of reconstituted carrot juice. *Food Hydrocolloids,* 20, 1190-1197.

Lucas, K., Maggi, J., Yagual, M. Creation of a tree tomato production, marketing and export company in the area of Sangolqui, Pichincha Province. Guayaquil, 2011, 140 p. Degree thesis (Commercial and Business Engineering). Escuela Superior Politécnica del Litoral. Faculty of Economics and Business.

Lutz, R. M. (2012). Can we talk about functional foods in Chile? *Revista Chilena Nutrición,* 39(2), 211-216.

Ly, M.H., Aguedo, M., Goudot, S., Le, M. L., Cayot, P., Teixeira, J. A., Le, T. M., Belin, J. M., Wache, Y. (2008). Interactions between bacterial surfaces and milk proteins, impact on food emulsions stability. *FoodHydrocolloids,* 22, 742-751.

Ma, Z., Boye, J. I., Fortin, J., Simpson, B. K., Prasher, S. O. (2013). Rheological, physical stability, microstructural and sensory properties of salad dressings supplemented with raw and thermally treated lentil flours. *Journal ofFood Engineering,* 116, 862-872.

Marquez, C., Otero, C., Cortes, M. (2007). Physiological, textural, physicochemical and microstructural changes of tree tomato *(Cyphomandra betacea S.)* in postharvest. *Journal of the Faculty of Pharmaceutical Chemistry,* 14 (2), 9-16.

Martin, D. A., Rico, D., Barat, J. M., Barry, R. C. (2009). Orange juices enriched with chitosan: Optimisation for extending the shelf-life. *Innovative Food Science and Emerging Technologies*, 10, 590-600.

Meng, Y., Rao, M. (2005). Rheological and structural properties of cold-water-swelling and heated cross-linked waxy maize starch dispersions prepared in apple juice and water. *Carbohydrate Polymers,* 60(3), 291-300.

Mewis, J., Wagner, N. Introduction to colloid science and rheology. In: Mewis, J., Wagner, N. Colloidal Suspension Rheology. Cambridge: Cambridge University Press, 2012, p. 1-34.

Meza, N., Manzano-Méndez, J. (2009). Tree tomato *(Cyphomandra betaceae* [Cav.] Sendtn) fruit characteristics based on aril coloration in the Venezuelan Andean Zone. *Revista UDOAgrIcola,* 9 (2): 289-294.

Mezger$_1$ T. G. (2006). The rheology handbook. Second Edition. Hannover, Germany: Vincentz Network, 300p.

Milani, J., Maleki, G. (2012). Hydrocolloids in food industry. In Valdez, B. Food industrial processes e Methods and equipment. Croatia: In Tech, 2012, p. 17-38.

Ministry of Agriculture and Rural Development. Municipal Agricultural Assessments (MADR-EVA). Bogotá: MADR, 2015, 181 p.

Ministry of Agriculture and Rural Development. Agricultural Statistics System - SEA. Statistical Yearbook of Fruits and Vegetables, 1992-2013. Bogotá: MADR, 2013, 304 p.

Mirhosseini, H., Ping, T. C. (2010). Effect of various hydrocolloids on physicochemical characteristics of orange beverage emulsion. *Journal of Food, Agriculture & Environment,* 8(2), 308-313.

Moelants, K. R. N., Cardinaels, R., Jolie, R. P., Verrijssen, T. A. J., Van Buggenhout, S., Zumalacarregui, L. M., Van Loey, A. M., Moldenaers, P., Hendrickx, M. E. (2013). Relation between particle properties and rheological characteristics of carrot-derived suspensions. *FoodBioprocess Technology,* 6(5):1127-1143.

Moraes, I. C., Fasolin, L., Cunha, R., Menegalli, F. (2011). Dynamic and steady-shear rheological properties of xanthan and guar gums dispersed in yellow passion fruit pulp *(Passiflora edulis Havicarpa). Brazilian Journal of Chemical Engineering,* 28(3), 483 - 494.

Mueller, S., Llewellin, E. W., and Mader, H. M. (2010). The rheology of suspensions of solid particles. *ProceedingsRealSocietyA.*, 466, 1201-1228.

Nascimento, G., Hammb, L., Baggio, C., De Paula, M., Lacomini, M., Cordeiro, L. (2013). Structure ofa galactoarabinoglucuronoxylan from tamarillo *(Solanum betaceum),* a tropical exotic fruit, and its biological activity. *Food Chemistry,* 141, 510-516.

Nundo, C. I., Tanga, J., Powers, J. R., Takhar, P. S. (2005). Rheological properties of blueberry puree for processing applications. *LWT.* 40, 292-299.

Nuñez, S., Scrofani, L. Determination of the efficiency of the dehydration process using the Demcom method in a medium crude oil. Puerto la Cruz, 2011, 80 p. Degree thesis (Petroleum Engineering). University of Oriente. School of Engineering and Applied Sciences.

Nwaokoro. O, G., Akanbi, C. T. (2015). Effect of the addition of hydrocolloids to tomatocarrotjuice blend. *JournalofNutritionalHealth & FoodScience,* 3(1): 1-10.

Ordonez, R., Cardozo, M., Zampini, I., Isla, M. (2010). Evaluation Ofantioxidantactivityand genotoxicity of alcoholic and aqueous beverages and pomace derived from ripe fruits of *Cyphomandra betacea Sendt. JournalofAgriculture andFood Chemistry,* 58, 331-337.

Osorio, C., Hurtado, N., Dawid, C., Hofmann, T., Heredia, F., Morales, A. (2012). Chemical characterisation of anthocyanins in tamarillo *(Solanum betaceum Cav.)* and Andes berry *(Rubus glaucus* Benth.) fruits. *Food Chemistry, 132*, 1915-1921.

Pabst, W. (2004). Fundamental considerations on suspension rheology. *Journal Ceramics Silikáty,* 48, 6-13.

Paquet, E., Alexandra, A., Lemieux, S., Turgeon, S. (2014). Effects of apple juice-based

beverages enriched with dietary fibres and xanthan gum on the glycemic response and appetite sensations in healthy men. *Bioactive Carbohydrates andDietaryFibre,* 4, 39-47.
Pegg, M. A. (2012). The application of natural hydrocolloids to foods and beverages. In: Baines, D., Seal, R. Natural food additives, ingredients and flavourings. Cambridge, UK: Woodhead Publishing Limited, 2012, 455 p.
Perez, P., Rocío. Application of microwaves in the treatment of water-in-oil (w/o) and oil-in-water (o/w) emulsions. Valencia: Spain, 2009, 290 p. Doctoral thesis. Polytechnic University of Valencia. Department of Communications.
Phillips, R. J. (2010). Structural instability in the sedimentation of particulate suspensions through viscoelastic fluids. *JournalofNon-Newtonian FluidMechanics,* 165, 479-488.
Pogribna, M., Freeman, J.P., Paine, D., Boudreau M.D. (2008). Effect of Aloe vera whole leaf extract on short chain fatty acids production by *Bacteroides fragilis, Bifidobacterium infantis and Eubacterium limosum. Letters in Applied Microbiology,* 46, 575-580.
Prabjone, R., Thong-Ngam, D., Wisedopas, N., Chatsuwan, T., Patumraj, S. (2006). Anti-inflammatory effects of *Aloe vera* on leukocyte-endothelium interaction in the gastric microcirculation of Helicobacter pylori-infected rats. *Clinical Hemorheology and Microcirculation,* 35, 359-366.
Pugh, N., Ross, S. A., ElSohly, M. A., Pasco, D. S. (2001). Characterization of aloride, a new high molecular weight polysaccharide from *Aloe vera* with potent immunostimulatory activity. *JournalofAgricultureandFoodChemistry*, 49, 1030-1034.
Quek, M. C, Chin, N. L., Yusof, Y. A (2013). Modelling Ofrheological behaviour of soursop juice concentrates using shear rate-temperature-concentration superposition. *Journal of FoodEngineering,* 118(4), 380-386.
Raeuber, H., Nikolaus, H. (1980). Structure of foods. *Journal of Texture Studies*, 11, 187198.
Rajasekaran, S., Ravi, K., Sivagnanam, K., Subramanian, S. (2006). Beneficial effects of Aloe vera leaf gel extract on lipid profile status in rats with Streptozotocin diabetes. *Clinical and Experimental Pharmacology and Physiology,* 33, 232-237.
Ramirez, Q. J., Aristizabal, T. I., Restrepo F. J. (2013). Preservation of blackberry by the application of an edible coating of aloe vera mucilage gel. *Vitae,* 20(3), 172-183.
Rao, M. A. Rheology of Fluid and Semisolid Foods: Principles and Applications. Second Edition. Geneva, NY, USA: Springer, 2006, 481p.
Restrepo, F. I. Conservation of strawberry *(Fragaria x ananassa Duch* cv. Camarosa) through the application of edible coatings of aloe vera *(Aloe barbadensis* Miller) mucilage gel. Medellín, 2009, 83p. National University of Colombia. Faculty of Agricultural Sciences. Department of Agricultural and Food Engineering.
Rivero, R., Rodríguez, E., Menéndez, R., Fernández, J., Del Barrio, G., González, M. (2002). Obtaining and preliminary characterization of an extract of Aloe vera L. with antiviral activity. *Cuban Journal of Medicinal Plants,* 7, 32-38.
Sagnay-Tanqueno, Monica. Comparative study of the nutritional potential of two varieties of tree tomato *(Solanum betaceum Cav.)* dehydrated by microwave at three powers. Riobamba, 2010, 120 p. Degree thesis (Pharmaceutical Biochemist). Higher Polytechnic School of Chimborazo. Faculty of Sciences.
Saha, D., Bhattacharya, S. (2010). Hydrocolloids as thickening and gelling agents in food: A critical review. *JournalofFood Science and Technology,* 47(6), 587-597.
Sahin, H., Ozdemir, F. (2007). Effect of some hydrocolloids on the serum separation of differentformulated ketchups. *JournalofFoodEngineering,* 81, 437-446.
Salinas, G., and Fuentes, F. (2012). Experimental evaluation of the behaviour of particle sedimentation velocity. *Revista Ingenierías Universidad de Medellln.* 2012, 11(20), 239-250.
Schramm, L. L. Emulsions, Foams, and Suspensions: Fundamentals and Applications. In: Schramm, L. L. Colloid Stability. Weinheim: Wiley-VCH, 2005, p. 117-152.

Shamsudin, R., Chia, S. L., Mohd, A. N., Wan, D. W. 2013). Rheological properties of ultraviolet-irradiated and thermally pasteurized Yankee pineapple juice. *Journal of Food Engineering,* 116(2), 548-553.
Sherafati, M., Kalbasi-ashtari, A., Ali-Mousavi, S. M. (2013). Effects of low and high acyl gellan gums on engineering properties Ofcarrotjuice. *Journal of Food Process Engineering,* 36(4), 418-427.
Sierra, G. A. Development of a prototype of an aloe vera *(Aloe vera barbadensis* Miller) and orange drink. Zamorano, 2002, 55 p. Undergraduate thesis (Agronomist Engineer). Panamerican Agricultural School (Zamorano). Department of Agricultural Science and Production.
Sigrid, S. M., Vidal, B. D. (2011). Functional foods enriched in Aloe vera. Effects Ofvacuum impregnation and temperature on the respiration rate and the respiratory quotient of some vegetables. *Proceed Food Science, 1,* 1528 - 1533.
Singh, A., Singh, A. K. (2009) Optimization of processing variables for the preparation of herbal bread using Aloe vera gel. *Journal of Food Science and Technology,* 46(4), 335338.
Song, K. W., Kuk, H. K., Chang, G. S. Rheology of concentrated xanthan gum solutions: Oscillatory shear flow behavior. *Korea-Australia RheologyJournal,* 18(2), 67-81.
Stading, M... 2011. Food rheology. *Rheology,* 1(2), 1-8.
Steenkamp, V., Stewart, M. J. (2007). Medicinal applications and toxicological activities of *Aloe* products. *Pharmaceutical Biology,* 45, 411-420.
Steffe$_1$ J. F. Rheological Methods in Food Process Engineering. Second Edition. Michigan, USA: Freeman Press, 1996, p. 418.
Stokes, J., Boehm, M., Stefan, B. (2013). Oral processing, texture and mouthfeel: From rheology to tribology and beyond. *Current Opinion in Colloid & Interface Science,* 18 (2013), 349-359.
Strickland, F. M. (2001). Immune regulation by polysaccharides: implications for skin cancer. *JournalofphotochemistryandPhotobiology,* 63, 132-140.
Suvitayavat, W. C., Sumrongkit C., Thirawarapan, S. and Bunyapraphatsara, N., (2004). Effects of Aloe preparation on the histamineinduced gastric secretion in rats. *Journal of Ethnopharmacology,* 90, 239-247.
Thompson, D., Harvey, A., Kazmi, M. and Stout, J. (1991). Fibrinolysis and angiogenesis in wound healing. *JournalofPathology,* 165, 311-318.
Torres, J. A., Tello, M. E., and Ostos, S. (2008). Development and optimization of an analytical methodology for the determination of sediment in cocoa-derived table beverage. *Revista Colombiana de Ciencias Químico Farmacéuticas,* 37(2), 177-190.
Urquijo, Jeaneth. Synthesis of magnetic nanoparticles and their implementation as ferrofluids. Medellín, 2007, 79 p. Degree work (Master in Chemical Sciences). University of Antioquia. Faculty of Exact and Natural Sciences.
Vandresen, S., Quadri, M. G., de Souza, J. A., Hotza, D. (2009). Temperature effect on the rheological behavior Ofcarrotjuices. *JournalofFoodEngineering*, 92(3), 269-274.
Wei, L., Chun-Cheng, Y., Hua-Feng, Z., Ru-Gang, Y. (2004) Preparation of aloe-herbs health beverage. *Food Science China*, 25(6), 207-209.
Williams, P. A., Phillips, G. O. Introduction to food hydrocolloids. In: Handbook of hydrocolloids. Florida, USA: CRC Press, 2009, 924p.
Yasar, Y., Togrul, H., Arslan, N. (2007). Flow properties of cellulose and carboxymethyl Cellulosefrom orange peel. *JournalofFoodEngineering,* 81, 187-199.
Yulianingsih, R., Maharani, D. M., Hawa, L. C., Sholikhah, L. (2013). Physical quality observation of edible coating made from aloe vera on minimally processed cantaloupe *(Cucumismelo L.). Pakistan JournalofNutrition,* 12 (9), 800-805.
Yusuf, S., Agunu, A., Diana, M. (2004). The effect of *Aloe vera* A. Berger *(Liliaceae)* on

gastric acid secretion and acute gastric mucosa injury in rats. *Journal of Ethnopharmacology,* 93, 33- 37.

Zheng, W. and Wang, S. Y. (2001). Antioxidant activity and phenolic compounds in selected herbs. *Journal OfAgricultural and Food Chemistry,* 49, 5165-5170.

Zhong, Q., and Daubert$_1$ C. R. Food Rheology. In: Kutz, M. Handbook of Farm and Food Machinery. Norwich, NewYork: William Andrew Publishing, 2007. p. 391-414.

Printed by Books on Demand GmbH, Norderstedt / Germany